Demand Response in Smart Grids

Pengwei Du • Ning Lu • Haiwang Zhong

Demand Response in Smart Grids

Springer

Pengwei Du
Electric Reliability Council of Texas
Taylor, TX, USA

Haiwang Zhong
Tsinghua University
Beijing, China

Ning Lu
Department of Electrical and Computer Engineering
North Carolina State University
Raleigh, NC, USA

ISBN 978-3-030-19768-1 ISBN 978-3-030-19769-8 (eBook)
https://doi.org/10.1007/978-3-030-19769-8

This Springer imprint is published by the registered company Springer Nature Switzerland AG
The registered company address is: Gewerbestrasse 11, 6330 Cham, Switzerland

Preface

Demand response (DR) is a reduction in load demand designed to respond to price signals or system dispatch instructions. DR programs have been used as resource options for balancing supply and demand by electric system planners and operators since the 1970s. Recently, resurging interests in DR technologies are motivated by advancement in information and communication technologies and by the wide deployment of sensors. The objectives of DR are also broadened to unfold the full potential of customer-owned distributed energy resources (DERs) for providing a full range of grid services.

In the future power grid, the penetration of DERs, such as energy storage, electric vehicles, roof-top photovoltaics, is expected to increase exponentially. Such modern power grids are facing many unprecedented challenges such as increased intermittency, operation uncertainties, and load consumption pattern shifts. On the other hand, there are limits to what can be achieved on the supply side due to the retirement and displacement of traditional generation resources.

To continue the past trend of investing on the supply-side alone to achieve reliable and secured grid operation will no longer be technically feasible or economically achievable. In recent years, researchers in the power and energy community have focused their efforts on developing a wide range of DR mechanisms to control controllable loads for providing generator-like services. As a result, the control mechanisms and capabilities of demand response have been greatly expanded.

This book is written to introduce to regulators, policy makers, practitioners, and researchers the history, ongoing developments, and future trends of DR programs. We first describe the load modeling principles and a few typical DR control methods. Then, we introduce the role of the wholesale market for monetizing the value of DR programs and the interplay between price formulation and load participation in energy and ancillary service market. After that, the design of retail market schemes for incentivizing DR with suitable price signals is discussed.

This book is divided into nine chapters.

Chapter 1: "Overview of Demand Response" presents an overview of DR, including the basic concept, history, control mechanisms, and the current status of load resources participating in the wholesale market. It also summarizes the lessons learned from those regions which have provided a steady revenue resource potential to DR participants.

Chapter 2: "Modeling Demand Response Resources" introduces the modeling of thermostatically controlled loads (TCLs), non-TCL controllable loads, and the base load. Unlike generators, end use electricity consumptions are heavily influenced by customer preferences so that the human factors are taken into account when modeling the working mechanisms of DR resources.

Chapter 3: "Basic Control Approach for Aggregated Demand Response Programs" describes load management algorithms, which are widely used to manage and coordinate load resources for providing grid services such as peak shaving, load shifting, load following, and frequency response. It discusses a few basic design principles that account for those distinct characteristics of load management programs in order to develop reliable, robust, and economical DR management systems.

Chapter 4: "Demand Responses in ERCOT" presents a review of current market design and the statuses of DR programs at the Electric Reliability Council of Texas (ERCOT). The performance requirements for DR participation and the benefits from deploying these load resources are illustrated through the real-world events.

Chapter 5: "Integrated Demand Response in the Multi-Energy System" reviews the state-of-the-art of Integrated Demand Response (IDR) which assists in integrating multi-energy systems. It covers the basic concept, the value analysis, and the key issues and provides a reference for future research directions on IDR programs.

Chapter 6: "Coupon Incentive-Based Demand Response" presents the formulation, assessment, theoretical analysis, and numerical illustration of a coupon incentive-based demand response (CIDR) targeting at retail customers who are equipped with smart meters yet still face a flat rate. Enabled by pervasive mobile communication capabilities and smart grid technologies, load serving entities (LSEs) could offer retail customers coupon incentives via near-real-time information networks to induce DR for a future period of time in anticipation of intermittent generation ramping and/or price spikes.

Chapter 7: "Distributed Real-Time Demand Response" discusses a real-time DR framework and model for a smart distribution grid. The model is optimized in a distributed manner with the Lagrangian relaxation method. Case studies demonstrate that the proposed method greatly outperforms the prevalent approaches, in terms of both efficiency and robustness.

Chapter 8: "Load Resources to Provide Primary Frequency Reserve Service" describes a co-optimization formulation of energy, inertia, and frequency response reserve in the day-ahead and real-time market where both primary frequency reserve from synchronous generators and fast frequency response reserve from load resources are procured in a cooperative way to meet the desired primary frequency response need.

Chapter 9: "Optimal Response of Residential House Load" presents a novel appliance commitment algorithm that schedules thermostatically controlled household loads based on price and consumption forecasts considering users' comfort settings to meet an optimization objective such as minimum payment or maximum comfort. The formulation of an appliance commitment problem is described using an electrical water heater load as an example.

The future of DR is promising as they offer many benefits to the grid and society as a whole. While this book covers the topics within this field as comprehensively as possible, more work is still under development. The intent of this book is not only to help a future grid to fully embrace the benefit of DR but also to encourage more people to contribute to this dynamic field and enable further exploitation of new revolutions for demand response. The future development of DR necessitates a collaborative approach to engaging all stakeholders and entities.

This book is written as an introductory book for engineers, economists, and policy makers at utilities as well as academic researchers to gain basic understanding of DR modeling and control, as well as pricing mechanism. It can serve as an excellent introduction for university students in electrical engineering at both undergraduate and postgraduate levels.

Finally, we acknowledge the innovative work contributed by others in this increasingly important area especially through IEEE Power & Energy Society Demand Response Working Group, and appreciate the staff at Springer for their assistance and help in the preparation of this book.

Taylor, TX, USA Pengwei Du
Raleigh, NC, USA Ning Lu
Beijing, China Haiwang Zhong

Contents

Chapter 1
Overview of Demand Response

1.1 Introduction

According to the Federal Energy Regulatory Commission, demand response (DR) is defined as:

"Changes in electric usage by end-use customers from their normal consumption patterns in response to changes in the price of electricity over time, or to incentive payments designed to induce lower electricity use at times of high wholesale market prices or when system reliability is jeopardized."

DR includes all intentional modifications to consumption patterns of electricity by inducing customers to alter the timing, level of instantaneous demand, or the total electricity consumption. As power systems are experiencing a period of rapid revolution, DR has been widely considered as a viable solution to improve the market efficiency and to enhance the grid reliability.

Bulk power systems are one of the most complex systems ever built, which consist of millions of components and a large number of large centralized generators. Maintaining a balance between supply and load is vital so that the frequency on an interconnected power system can be regulated within a normal range. Historically, this responsibility was primarily performed by the generation supply while the load was not actively participating in the balancing process. This is often characterized as a "generation-follows-load" paradigm where the generation units are dispatched in response to the anticipated load changes. In addition, a certain amount of expensive reserves are carried over by generators for unexpected generation outages. If there is a severe shortage in generation, excessive power capacity needs to be brought online quickly. Otherwise, involuntary load shedding may be triggered. This power deficit may often be accompanied by a skyrocketed energy price. Rather than relying on generation resources to balance load changes, controllable load demand can alternatively be deployed to achieve the same objective. From a balancing perspective, the curtailment of load resources can emulate an increase in generation power output in such a way that the corresponding effect on restoring power balance will be the

P. Du et al., *Demand Response in Smart Grids*,
https://doi.org/10.1007/978-3-030-19769-8_1

same. This type of intentional changes in load is induced as a consequence of a wide range of actions which consumers may take, collectively called DR. Since invented in the 1970s, DR applications have been tested and deployed worldwide. Thanks to recent investment in smart grid technologies, especially massive installation of smart meters, the potentials of DRs are being fully exploited in a large scale, which makes DRs to become the increasingly valuable resources to improve both efficiency and reliability of power systems.

DRs differ in various aspects from their generation counterparts, which may pose unnecessary regulatory, market and technical barriers to the integration of DRs into the grid if not addressed properly. First, DRs are more fast-acting in changing their power demand. Due to the ramp rate limitations, large generation units cannot change their power output instantaneously when needed. The absence of flexibility, which is the ability of a power system to respond to change in demand and stochastic supply, is one of major constrains to integrating a high amount of intermittent renewable resources. In contrast, DRs are much more responsive to the grid need as they allow a larger ramp rate. In some circumstances, DRs are controlled as block response, not as smoothly as the proportional response of generators. Second, the operational characteristics of load resources may not be well known, and sometimes it is not trivial to build their analytical models due to the heterogeneous and diverse nature of load resources participating in DR programs. Third, operation constraints of DRs may be much more complicated compared to the generation supply. For example, ideally, the adjustment and shift of load demand should not cause significant impact over the energy service, such as heating or cooling to maintain a given indoor temperature range. In this regard, the activation duration of DRs may be limited, and end-users' preference also needs to be considered when deciding when and how to deploy DRs. Forth, when a large number of small-size load resources are aggregated to provide grid services, their modeling and controlling are challenging.

The underlying objective of DRs is to actively engage customers in modifying their consumption in response to pricing signals or grid need. There are three types of demand response: emergency demand response, economic demand response, and ancillary service demand response. Emergency demand response is employed to avoid involuntary service interruptions during times of supply scarcity. Economic demand response is to allow electricity customers to curtail their consumption or postpone their energy usage in an exchange for an economic return, which is expected to be greater than loss in productivity or the sacrifice in customer convenience. Ancillary service demand response can be deployed quickly and precisely to ensure the secure operation of the transmission grid.

1.2 History of Demand Responses in North America

The development of DRs went through several phases in the past. Historically, its advancement pace was largely driven by the evolution of communication and control technologies, market deregulation and regulatory changes, and incentive to DR participations.

DRs first gained momentum in the 1970s as the cost of producing power increased while residential demand grew quickly. Natural gas shortages and oil embargo (1973 and 1979) drove up the cost of energy. Moreover, fears of meltdown (1979 Three Mile Island) deterred further investment in nuclear power. National Energy Act of 1978 slowed growth of demand and restructured how utilities purchased power. Public Utilities Regulatory Policy Act (PURPA) required purchases of energy from Non-Utility Generators (NUGs). All of these led to increased usage of DRs and coincident peak rate design, especially for commercial and industrial customers. These schemes were known as Demand Side Management (DSM) or a combination of DR, energy efficiency measures, and consumer education. However, many argued that DRs would hinder energy conservation efforts since DRs shift consumption rather than reducing demand.

Between 1975 and 1976, the Federal Energy Administration initiated a number of pricing experiments to assess impact on load patterns. Peak reductions between 0 and 42% were observed. Off-peak increases were negligible so that daily consumption was reduced between 0 and 6%. These experiments highlighted possible advantage of time-of-use pricing.

In the early 1980s, growth of DRs slowed. Oil prices collapsed and natural gas was deregulated. Demand growth slowed, leading to excess generation capacity. State regulators widely instituted "least cost planning." If DR was cheaper than generation costs, utilities were required to implement DR programs. Recovery rates for nuclear energy were significantly reduced, and incremental electricity rates were generally greater than cost of DR program. This led to numerous subsidy and incentive programs. In general, exacerbated excess capacity drove up regulated rates.

In the late 1990s to mid 2000s, DR resources decreased significantly (by 32% from 1996 to 2004). California market was deregulated in 1996, followed by many other states, and this led to regulatory uncertainty. During this period, technology was not mature to take advantage of the potentials of DRs as incremental cost of DRs still far exceeded the incremental cost of new generation.

In the late 2000s and 2010s, DR experienced a resurgence and the success of DRs has been fully unfolded. American Recovery Reinvestment Act (ARRA) invested ~$4.5 billion for electrical infrastructure. New technologies and rate structures were deployed and tested, especially for control and response of residential load. This led to a large-scale installation of smart meters, which can measure energy usage at a high time resolution of 15 min and bring the real-time energy price to end-users. In March 2011, FERC ruled "DR is equivalent to generation and supply side resources and thus deserves equal compensation in wholesale markets." This was formalized in FERC order 745. The regulatory change further accelerated the deployment of DR programs as more incentives are provided to demand response resources.

1.3 Benefits of Demand Responses

With their capabilities expanded over time, DRs are capable of performing a wide range of services to the grid, from peak load reduction, congestion management, assistance in emergency conditions to the provision of ancillary services.

1.3.1 Peak Load Reduction

DRs provide consumers an opportunity to play a significant role in the operation of the electric grid by reducing or shifting their electricity usage during peak periods in response to time-based rates or other forms of financial incentives. Activation of DRs during peak conditions can bring about tremendous reliability benefits and financial savings. Power systems exhibit nonlinear characteristics which result in variable efficiencies. Under heavy loading conditions, losses are higher due to nonlinear quadratic dependence on currents. During peak times, less efficient "peaking units" may need to be run. As a result, systems with high peak load characteristics are less efficient and more expensive to operate.

More noticeably, capital investments are driven up by this peak load. Electric power systems are designed to support highest peak during the peak day of the year, which is due to coincidental usage and is driven by the end-users. As the utilities need to keep a balance between load and generation at every instant, large central generation plants have to be installed and dispatched. In addition, from a reliability perspective, a sufficient reserve margin must be maintained, which requires more idle generation capacity. Because the load is not an active participant in balancing, the whole electricity infrastructure has been designed and operated to account for this. As a result, costly infrastructure has been constructed to accommodate peak system conditions.

Figure 1.1 depicts the representation of hourly loads as a fraction of peak load in Western Electricity Coordinating Council. When sorted from highest to lowest, it yields the efficiency of utilization of electricity infrastructure. It clearly shows that 25% of distribution and 10% of generation assets (transmission is similar), worth of hundreds of billions of US dollars, are needed only for less than 400 h each year. Electrical infrastructure is definitely underutilized for the majority of the time.

Engaging DRs at peak times could yield some of the most significant benefits while affecting the end use consumers the least. A report prepared by the Department

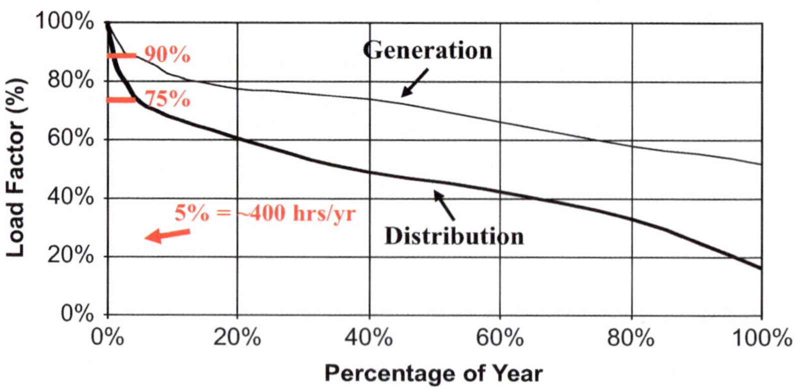

Fig. 1.1 Sorted load factor curve

of Energy in the USA identified and quantified the national benefits of demand response and made a recommendation on achieving specific levels of such benefits. The report estimated that in 2004, potential demand response capability equaled to about 20,500 megawatts (MW), 3% of total US peak demand, while actual delivered peak demand reduction was about 9000 MW (1.3% of peak), leaving ample margin for improvement.

Deploying DR during the highest peaks not only yields the greatest benefits, but also comes at low costs and less requirement for the communication infrastructure needed to enable such a deployment. In this regard, this category of DR program has found greatest success since the onset of development of DRs.

1.3.2 Congestion Management in Distribution Network

Congestion management is a common problem in power systems and becomes more pronounced in deregulated and competitive markets. Distribution lines are often driven close to or even beyond their thermal limits in order to satisfy the increased electric power consumption or deal with increase of the unplanned power exchanges. If the power flows were not controlled, some lines located on particular paths may become overloaded and this phenomenon is called congestion. There are a number of reasons why congestion management in distribution networks is difficult. First, the distribution networks are operated with radial topology, thus leaving no alternative path to alleviate the overloaded power flow for outages. Second, the overloading of distribution lines may not coincide with the peak load hours, as load resources connected to distribution networks are diverse and varying. Thus, it is hard to estimate the proper size and capacity of circuits needed in the planning stages. Third, the investment on the distribution networks is lagging behind the need in North America. It is estimated that the utilities in the USA need to spend billions of dollars to upgrade the aged distribution infrastructures.

The congestion management in distribution systems is also complex and could lead to severe interruption of load if not properly handled. Basically, the alleviation of congestion can fall into two categories, depending on how much the cost may be incurred. The first category is to adjust operation of transformer taps/phase shifters, power electronics-based compensation devices, or reconfigure the system. The second category is to re-dispatch the generation or apply load curtailment combined with re-dispatch. However, neither of them is efficient without incurring a high cost. The first category of actions may only have limited effect over removing overloading conditions, while the cost of relieving congestion would be prohibitively high for the second category of congestion management schemes.

DRs can solve the congestion problems in an efficient way. Rather than directly shedding load, congestion management through DRs focuses on rescheduling or postponing load demand to reduce the congested power flow. These load resources selected are either flexible or have thermal inertia. They are particularly suitable for doing this type of jobs as the shutdown for a short period of time could not even be noticeable to end users. To facilitate the coordination of actions needed from a large number of load resources, a local market could be introduced. Each load willing to participate in this market can submit a bid curve, indicating their cost of providing congestion management service. All bids are evaluated relative to their effectiveness in alleviating congestion. The benefit of establishing such a local market is that multiple constraints and objectives can be considered under this framework. As the participation in congestion management is cost based and voluntary, disputes could be largely avoided compared to the traditional load curtailment scheme directed by operators.

1.3.3 Ancillary Service (AS)

Ancillary Services are services necessary to support the transmission of energy to loads while maintaining reliable operation of the transmission systems. These operating reserves serve as additional capacity to keep a balance between load and generation in case that a generating unit goes offline, load is higher than anticipated, or another problem emerges. The ancillary service products can be generally characterized by their response time, duration of the response, the time to fully response, and how often the product is called (whether they are deployed continuously or only under contingency conditions)—see Fig. 1.2.

Ancillary service markets vary substantially across regions, but often include a "regulation" product requiring short time-scale adjustments to load within a real-time dispatch interval, one or more "spinning" reserve products requiring responses

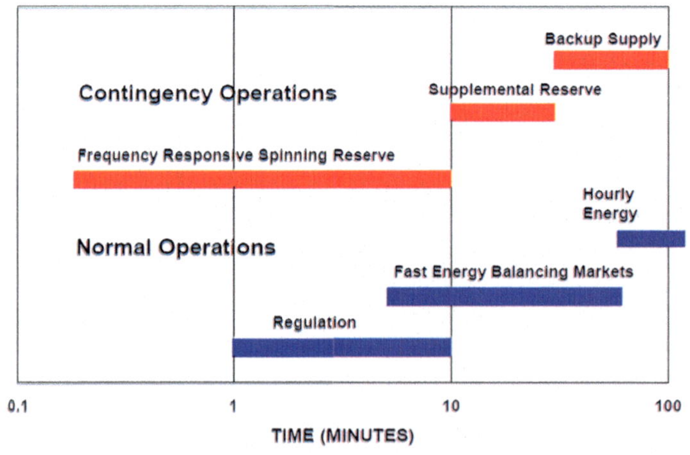

Fig. 1.2 Response time and duration that characterize ancillary services (Source: US DOE, "Benefits of Demand Response in Electricity Markets and Recommendations for Achieving Them: A Report to the United States Congress Pursuant to Section 1252 of the Energy Policy Act of 2005", Feb. 2006)

at short notice (seconds to 10 min) during a contingency, and "non-spinning" reserve products requiring responses with a longer lead time (30 min). In the event of a contingency, fast-responding products will activate first, helping to support system frequency. As products with longer lead-time activate, they help restore the system to its original state.

The ancillary services were predominantly provided by generation supply in the past mainly for two reasons. First, significant market and regulatory barriers existed for DRs to participate in ancillary service markets. Second, DRs lacked capabilities to deliver the obligated response when called upon before recent technology advancement in communication and control was developed. Due to the variable and uncertain nature of load, DRs may encounter difficulties in precisely following ancillary services deployment instructions without a dedicated control system.

On the other hand, controllable load resources represent a significant and largely untapped resource for supplying reliability services to electric power systems. To only allow generators as the sole resources to provide ancillary services could lead to a scarcity condition if the generation supply capacity is already tight, which could jeopardize the system reliability. In addition to the improved grid reliability, price volatilities and market inefficiency could also be mitigated if ancillary service markets are open to DRs. As the value of DRs in provision of ancillary services is being increasingly recognized, new communication and control techniques have been developed to improve the controllability of DRs, through either individual or aggregated response.

To eligibly participate in ancillary service market, DRs must meet the minimum qualification requirements for their size and maintain a level of responsibility as scheduled. DRs are best suitable for providing contingency reserve to restore the power balance when large generators are tripped offline. If activated by under-frequency relay, the response of DRs delivered as contingency reserve is much faster than generator governor actions, which is more beneficial to improving the system reliability. DRs can also provide regulation services. As the load exhibits variations at different frequencies, in order to maintain the generation-load balance, generators must change their output continuously. DRs can potentially be used to "smooth" the fluctuations by providing regulating services. The deployment of regulation services is achieved through automatic generation control (AGC) system, which issues deployment instructions every few seconds. Therefore, the control system of DRs contracted for regulation services needs to be upgraded so that they can receive and respond to these AGC commands. The requirement for DR's participation in supplemental reserve is less demanding, which consists of a 10-min or 30-min ramp rate limit. Among those ancillary service markets, contingency reserve is the most attractive to DRs as it provides a largest market opportunity for participation and the payment for such a service is also highest.

The ability for DRs to provide ancillary service is largely affected by the flexibility of the underlying loads, their availability, and the aggregation scheme. Flexibility is a function of the load regarding whether the load can be temporally interrupted while meeting its operating characteristics and constraints. The availability of DRs is the load consumption pattern which limits the maximum capacity of

DR to be offered in ancillary service. The aggregation scheme influences the cost of DR's provision of ancillary service and determines the performance of delivered demand responses. All of these contribute to the difficulty for DRs in maintaining the ancillary service responsibility once it is awarded, and could hinder their abilities to compete with conventional supply-side resources.

In an existing market, the price of ancillary service is cleared as the bid cost of the marginal unit providing such a reserve. However, unlike generators, the load demand is constrained by multiple factors including economics, flexibility and availability, and a need to hedge against energy price spike. In some cases, the primary goal for load resources is still to provide energy service to the underlying processes rather than making extra profits from ancillary service market. Thus, it is much more complicated and harder to submit a bid curve for DR which reflects its actual cost of providing ancillary services. For this reason, DR often acts as price-taker, which leads to a reduced incentive to DR for their participation in ancillary service market.

1.3.4 Emergency Conditions and Prevention of Blackouts

While electricity is essential to the health and prosperity of our society, our grid is also susceptible to large-scale blackout as bulk power grids have evolved into one of most complex and vulnerable systems invented by humans. The 1996 and 2003 blackout events in North America have incurred a financial loss in excess of billions of dollars.

The mechanism of causing cascading failure is still not well understood. Power grid is designed and operated with a redundancy so that it can withstand a loss of any one or two components (N-1 and N-2 security). However, blackouts which have occurred were not envisioned or beyond the design basis of the system. Moreover, the speed of cascading failures is too fast, leaving operators no time to make decisions to intervene the process. All of these lead to lack of efficient means for blackout prevention.

DRs have proven tremendously successful for the utilities in managing a peak-demand crisis. As DR program is growing quickly in popularity, in the absence of other means to avoid blackouts, DR can play a growing role in preventing and mitigating blackouts. Historically, DRs were recruited in direct load control program to assist in an emergency situation. When the power system is in stress and a high risk of insecurity is expected, these load resources can be called upon at the discretion of system operators. In return, DRs are compensated for the capacity made available for deployment during emergencies.

Another approach is to install automatic load shedding devices that respond appropriately to limit the consequences or even stop a cascading failure sequence. These load shedding actions usually respond to a frequency excursion or the system command, and the resulting action of this control serves to help bring the system back into equilibrium. As this relatively small amount of load is shed faster and earlier to restore the balance, the cascading failures could be stopped at the initial stages before the whole blackout process is not reversible. These controls can be autonomous and isolated, making them very secure and robust. With this approach, load reduction is controllable and transparent to the end-use customers. Moreover, a priority list can be used for deploying load resources according to their flexibility and cost of load interruptions.

Opportunity also exists to identify innovative technologies to develop more effective methods for blackout prevention. Load shedding events typically impact entire circuits, with all customers on the circuit losing power. However, with increasing deployment of advanced metering infrastructure (AMI), utilities can significantly improve the precision of load shedding as AMI provides greater granularity in control. Load resources can also be aggregated together to eliminate some uncertainties in the action of the single load. When the aggregation of load resources meets a minimum size requirement, it can be modeled and monitored in the energy management system (EMS). By aggregation, it increases the potential reward of expanded DR participation, while controlling risk to end-users.

1.3.5 Transactive Control with Continuous Response to System Imbalance

DRs can continuously respond to system imbalance based on a bidirectional "transactive" response approach, which was invented by researchers at Pacific Northwest National Laboratory (PNNL). This represents a paradigm shift to how the power grid is operated. Transactive energy systems provide the control and coordination required to actively engage customer-owned and third-party assets to provide flexibility through transparent, competitive means. In contrast to event-based DR response, transactive response approach allows DR to adjust their power consumption on a continuous base and in a precise way.

According to GridWise Architecture Council, transactive energy is "a set of economic and control mechanisms that allow the dynamic balance of supply and demand across the entire electrical infrastructure using value as a key operational parameter." The term "value" here basically corresponds to economic and market-based standard values of energy. Essentially, transactive control establishes locational, 5-min transactive "price" signals to incentivize consumers to align their demand with power supply conditions. These "price" signals are synthesized to reflect the actual grid need. As they are conceptually different from location marginal price (LMP), these "price" signals are better understood as a time-varying control command created to coordinate operations of supply and load resources. These self-controlled changes in generation and load resources are in a great demand as the fluctuations in the power produced by intermittent renewable resources are rapidly increasing.

Transactive energy promotes a decentralized environment for distributed energy nodes as opposed to the traditional hierarchical grid structure. By design, the key component in transactive energy is a transactive node, which is a node equipped with an agent that communicates with other agents and makes automated decisions. Such a node will first submit a bid for its schedule which it is willing to do while considering its internal operational constraints. The next step is that all agents communicate at different levels to settle down at a "transactive" price, which then is broadcasted to each node. Upon the receipt of the price, each node will determine its own schedule as the intersection between the transactive price and its bidding curve. These procedures may be performed in an iterative way if multiple layers are involved. The schedule of transactive energy can be determined either day-ahead, hour-ahead, or in real time.

Transactive control has the following features. First, transactive control can be in force at time scales from minutes to hours, whereas typical demand response takes place at time scales of hours or days. Second, transactive energy is an autonomous response and based on a decentralized decision making framework so that it is scalable to coordinate millions of devices. Third, these devices are "controlled" based on economic incentives rather than centralized commands. As transactive control leaves the decision making up to the consumer, the participation of devices is voluntary. Fourth, both supply-side resources and demand-side resources are eligible to become a transactive node.

This transactive control DR system was first tested in field in Olympic Peninsula Smart Grid Demonstration Project where over 100 residential household loads participated. The experiments proved that the customers were able to and willing to respond to real-time price information in return for an economic gain. More significantly, the system demonstrated a remarkable ability to control system demand peaks and create a stable and predictable load profile for the utility. Given the success in Olympic Peninsula Smart Grid Demonstration Project, transactive control was further demonstrated in a much large scale in the Pacific Northwest Smart Grid Demonstration project which costed $179 million and lasted for 5 years. The Pacific Northwest Smart Grid Demonstration project consists of 27 subregions that can exchange information with one another. Each subregion negotiates with neighboring subregions to determine the quantity of import or export and the corresponding cost. The experiences in Pacific Northwest Smart Grid Demonstration project show that transactive control is promising in utilizing the flexibility from a large number of distributed resources and intelligent devices.

1.4 Motivations and Barriers for Implementing Demand Responses

There are various reasons why the consumers choose to participate in DR programs. They can gain economic advantage on their bills as they are offered a lower energy price or payment for their participation in DR programs. The deployment of DR can help avoid blackouts and potential cost increases so that the benefit can be socialized cross the whole grid.

From utilities perspective, the motivations for the implementation of DRs are apparent. DRs can potentially avoid emergency situations by providing contingency response. The long-term, capital intensive capacity investments can be deferred. Deployment of DRs can also reduce exposure to volatile energy costs and wholesale market prices.

However, the barriers for both consumers and utilities to engage in DRs are still in existence. Consumers will hesitate to participate in DRs if they lack understanding about DRs. If not well educated, they may be concerned with the possibilities to change the way they live or the risk of being exposed to price spikes. Thus, consumers may think that it is too difficult and cumbersome to use DRs. Consumers also need to know how much money can actually be saved from participation in DR programs. Studies indicate that $10–15 per month is the minimum amount to make DRs worth the effort.

From utility side, it may be difficult or costly to implement DR programs if DRs become dispatchable. Utilities may be worried that DR will not respond as expected so that they have to buy excess capacity to cover unresponsive DR resources. Another concern which utilities may have is that consumers could "game" the system to make money at the expense of the system.

1.5 Demand Response Control

To bring about a change in load, utilities may signal demand requests to their customers in a variety of ways, including simple off-peak metering, in which power is cheaper at certain times of the day, and smart metering, in which explicit requests or changes in price can be communicated to customers. In advanced DR applications, there will be a form of control input. The input is often a control signal in the form of a price signal, which is sent by the utility or the system operator. While the price signals are one type of control signals, other variables could also potentially be used. Upon the receipt of the control signal, the customers may adjust power demand by postponing some tasks that require large amounts of electric power, or may decide to pay a higher price for their electricity.

Once a request for the change is explicitly or implicitly sent to customers, there are two ways to implement a DR control. In the first implementation, a price signal may, or may not, be sent to each residence. Customers are informed that prices vary according to some schedule. It is up to the customer to determine how, and if, they should change the way they are currently using electricity. In a second way to implement a DR control, the utility, or a third party aggregator, sends a price signal to the customers. The price signal is received by devices such as automated thermostats that respond to the price signal based on internal set points as well as set points determined by the end-use customer. In this way the end-use customer sets their preferences and the hardware implementation automatically executes the routine operations based on pre-programmed logic.

DRs can also be classified into different categories depending on how DRs are deployed. In incentive-based programs, users are offered monetary incentives and agree to reduce load to help maintain system reliability or to avoid high generation costs. Interruptible load and direct load control fall into this category in which load serves as a capacity resource and participants commit to load reduction by prespecified levels when system contingencies arise. In price-based programs, time-varying price signals are provided to induce consumers to reduce energy usage during high-price hours. These include time-of-use tariffs, critical peak pricing tariffs, critical peak rebate tariffs, and real-time pricing tariffs. The third category is controllable load as resources, which offer a bid into the wholesale market and could set the price. Each control has its pros and cons and is suited to different DR goals. More details for these different kinds of DRs are provided as follows.

1.5.1 Interruptible Loads

Interruptible loads are the most direct form of DR. Usually utilities sign an agreement with large industrial customers. Utilities pay for the right to turn off very large loads (e.g., an aluminum smelter) with a predefined amount of notice. In exchange, the customer typically receives a reduced rate. In an interruptible load program, participants are subject to load interruption during system contingencies. However, interruptible loads tend to have high dropout rates when they are called too often.

1.5.2 Direct Load Control (DLC)

In a direct load control scenario, the utility has a direct control path to the appliance controller such as heating, ventilating, and air conditioning (HVAC), water heaters, or pool pumps. Once the permission of the end user is obtained, the utility can control the end use load within the limits of a contract. The command sent to load is either an on/off control similar to interruptible load or a variable duty cycle signal, i.e., HVAC can be controlled to be cycled on and off via a switch on the compressor, or by adjusting room temperature set points via a smart thermostat. The end user is then given a monetary rebate for allowing this type of control. Utilities prefer this type of systems because of certainties in DR deployment such as when DR will be triggered and how much load will respond.

1.5.3 Time of Use (TOU)/Critical Peak Pricing (CPP)/Critical Peak Rebate (CPR)

TOU, CPP, and CPR share a common feature as tiered rates since all of them typically change in blocks of hours in a predetermined way. Within the framework of TOU, the price is varied throughout the day (or year) to achieve load reduction at predetermined times. TOU is typically a 2- or 3-tier price rate (2–3 price levels per day) and has seasonable variations for summer and winter. The changes in load demand can be induced as a response to the tiered rate in two ways. Either the system informs the customer of electricity costs at different amounts at different times and leaves the end user adjust their usage, or the appliances will respond in a pre-programmed manner (often immediately and automatically). CPP is similar to a TOU except that power companies are entitled to call critical events during a time period of high wholesale market prices and/or system emergency conditions. A critical event lasts for a limited number of hours, within which electricity prices increase substantially to incentivize users to reduce energy usage. When critical events are called, the time and duration of the price increase can either be predetermined or vary based on how much load needs to be reduced in the events. CPR is analogous to CPP, except that during a critical event, electricity prices remain the same while a user is refunded for a predetermined rebate. The billing factor is a user's usage reduction relative to what the utility's expectation is.

1.5.4 Real-Time Price (RTP)

Until recently before smart meters are massively installed, a majority of customers (i.e., residential and small commercial) overwhelmingly only have usage meters which simply accumulates the usage over time. As such, these customers can only be billed for the electricity they use based on flat rates or tiered rates. Such electricity rate designs do not convey the variability over time in the cost to produce electricity. Using smart meters, utilities are now capable of recording electricity usage on a much more frequent basis (e.g., every 15 min), enabling utilities to introduce new types of pricing programs that better reflect these differences over time in the cost to produce electricity. An RTP is similar to a TOU price except that the price can change every 5 min (or 15 min or h). Because it changes so often, this increases complexity of DR control and requires utilities to upgrade infrastructure. However, because of the short time scale, it can be more effective and more accurate than a TOU to represent the true cost of consuming electrical energy, and also known as Dynamic Pricing. An RTP can be the actual market price cleared in energy market, or a "price" signal in transactive control to guide the responses toward a new equilibrium. In the latter case, the price signal sent to the end user may not be the actual cost of producing power, but indicates the relative cost of power at a given time. By using a price signal, the end users do not have to understand the complexity involved in operating a power system. In summary, these price-responsive customers participating in RTP react to prices by adjusting their demand, but without bidding into the wholesale market, i.e., electricity consumers tend to reduce their consumption if the prices rise.

1.5.5 Controllable Load as Resources (CLR)

Controllable load as resources (CLR) offer load reductions as a supply-side resource (comparable to generation offers) and can participate in all three types of wholesale markets: energy, ancillary service, and capacity. Basically, those CLR customers need to respond to instructions by the utilities, aggregator, or the system operator. In return, customer receives an explicit payment for curtailing load or the capacity made available for dispatch. In contrast to dynamic pricing-based DRs which only react to market prices without bidding, this kind of DRs is fully integrated into the wholesale market through demand bidding, which reflects a minimum amount of payment needed for CLR to curtail demand. Additionally, CLRs are allowed to set the market price. In order to participate in the whole market, CLRs should also be capable of meeting notification and response time requirements and following the dispatch instructions received.

1.6 Demand Responses in Wholesale Market

The US electricity industry began to transition toward the wholesale power market in the 1990s following FERC order 888, which mandated open access to transmission systems. The restructuring established regional wholesale market, and thus produced more incentives to attract load resources to participate in market-based demand programs than traditional utility-run demand side management programs. DRs currently provide an array of services to support the daily operations of wholesale market. Meanwhile, the amount of DRs is steadily increasing in those markets which are open to more participants from both supply and demand side.

The degree to which DRs are integrated into a market also depends on the ability of load resources responding to price or dispatch instructions. Larger electricity consumers especially benefits from the deregulation since they may be able to quickly respond to short-term price signals and adjust their consumption accordingly. Smaller customers are still faced with market barriers and overwhelming cost to meet metering and response requirements so that their participation in the wholesale market is limited. For this reason, most of the small customers often pay retail prices that do not follow short-term movements in the wholesale price. On the other hand, the innovations in technologies have opened up new routes for small customers to participate in the market. As the participation of DRs in the wholesale market is greatly expanded, their impacts on the grid reliability and market efficiency are also broadening.

1.6.1 Overview

In North America, DRs are allowed to participate in all of the three markets, namely energy market, ancillary service market, and capacity market. In the following, we present a high-level summary for the market rules and the recent development of integrating DRs in these markets.

1.6.1.1 Demand Response in Energy Market

Load resources can simply be responsive to price signals without system operator control. However, dispatchable and controllable load resources may take advantage of the wholesale energy market and receive more payment from load curtailment if scheduled by the system operators. When DRs actively participate in the wholesale energy market, they will be dispatched for economic reasons. Those resources need to bid their demand reduction directly into either the day-ahead market or the real-time market. If the bid is accepted, the resource will be dispatched by the system operator and receive the energy market price as a payment.

Several markets already allowed customers to participate directly in the wholesale energy market by submitting a schedule of quantity and price bids for demand-side dispatch, similar to how generators submit quantity and price offers for supply-side dispatch. While this option was made available to DRs in some jurisdictions, customers generally lacked an interest in participating in energy market due to a few reasons. First, the costs to purchase real-time telemetering equipment and the loss of a customer's operational flexibility when its consumption is controlled by the system operators are not justified compared to the incremental value from demand response. Second, the wholesale energy price may not be volatile, and this presents an uncertainty to the revenue paid to DRs when they are participating in the whole energy market. Third, it may be not without any controversy to determine the actual response of DRs when dispatched, which requires to establish a hypothetical baseline level that would have been consumed if the resource had not been dispatched.

Allowing DRs to be dispatchable rather than price-responsive enhances economic efficiency and supports supply adequacy. Ideally, under scarcity conditions, dispatchable demand reductions would clear the market and set energy prices at customers' willingness-to-pay. In doing so, market rules permit determining the system price while taking proper account of DRs, which could result in efficient price formation that supports efficient operations and investments in the long term.

1.6.1.2 Demand Response in Ancillary Service Market

DRs that have the ability to curtail their load demand quickly or to follow a basepoint dispatch instruction may participate in AS markets. Examples of AS include regulation, responsive reserves, and non-spinning reserves. These ancillary service products are intended to help balance supply and demand at short time scales, including following a contingency.

Future power grid cannot be operated reliably without carrying a sufficient amount of AS. Traditionally, AS has been provided exclusively by generators where unloaded generation capacity is reserved and dispatched in a short notice in real time when needed. For this reason, AS procurement mechanisms and market rules have been specifically designed to accommodate the physical characteristics of generators, rather than the essential qualities necessary to perform AS functions. However, incorporating DRs in AS markets definitely increases market competition, reduces the service cost, and improves the market efficiency.

Several market barriers need to be removed in order to allow DRs to compete with generators in AS markets. The most significant barrier is caused by the definition of AS products, explicitly excluding certain classes of load resources for participation. By the market design or qualification requirements, the attributes of performance (e.g., minimum resource size) and the required enabling infrastructure (e.g., telemetry) for load resources may hinder DRs from participating effectively in AS markets. In addition to technical issues, other barriers which could impact the DR potential are those requirements that increase the cost of supplying AS, decrease the revenue potential for AS providers, or limit enrollment of retail customers.

Load resources are able to provide AS which generators can do. In some circumstances, loads can perform better than their generator counterparts. For example, when load resources are curtailed automatically using an under-frequency relay, their responses are extremely faster than the generator's governor actions so that the new operational equilibrium can be restored quickly following the contingencies. Experiences at Electric Reliability Council of Texas (ERCOT) show that DRs can provide up to 60% of responsive reserve, and their fast response brings tremendous benefits to the system reliability when the system inertia is low.

Market operation experiences show that a larger amount of DRs have entered the AS market rather than the energy market. In contrast to the energy market where DRs are compensated based on their response, AS providers receive an availability payment for the capacity. Moreover, energy market prices exceed DR strike prices only rarely so that the revenue in energy market, which is considered to be volatile and uncertain, is insufficient to incentivize a large number of DR providers to enter the market. In comparison, available revenues from being an AS provider is sufficient and able to be captured with enough certainty to meet return on investment levels for fixed and variable enabling infrastructure to participate in the AS market.

1.6.1.3 Demand Response in Capacity Market

The wholesale energy markets have been running very successfully since their inception, but not without a challenge. One of these challenges is to ensure resource adequacy, an important aspect of the power system reliability. A well-functioning power market should be able to send appropriate pricing signal to market participants so that they wisely decide whether to invest on the new generation resources.

However, it is often the case where sufficient generation capacity has not been built to meet annual peak demand, which is called resource adequacy problem. In some regions, extremely high energy price cap is used to alleviate the scarcity conditions. However, new generation resources are still not constructed until the reserve margin is becoming very low, at a point threatening the system reliability. In other regions, capacity market is sought for procuring sufficient capacity to ensure system reliability. While the market rules for capacity market vary, their fundamental principle is the same, i.e., capacity market products are intended to meet traditional planning reserve requirements (or targets) that support reliability, especially during super-peak load conditions.

A capacity market is featured by the forward-looking capacity auction, which is held annually or seasonally, at a much longer leading time than AS market. In such an auction, regional system operators solicit bids from resources, who agree to be available during the expected peak hours, to meet estimated future peak demand. Resources that are cleared in the auction receive a stream of market-based compensation determined by the auction price and the cleared capacity. This represents additional incentive to new resources if these capital-intensive resources will be built, on the top of the revenue which new resources receive in energy or AS market.

DRs can assist in ensuring resource adequacy by curtailing load on a short notice when a capacity deficit is anticipated. Similar to the generation resources, DRs bid in a set amount of load that can be curtailed during a capacity shortfall, especially during expected peak hours. In doing so, DRs provide system operators more choices with the ability to reduce demand when needed. In addition, DRs cleared in capacity market generally receives an availability payment for the cleared capacity over years and may or may not receive an additional payment for curtailments actually called.

Recent capacity market developments have focused on participation rules for DR providers to address one of the important issues, i.e., how to reconcile the different requirements for generators and DR resources in capacity market. A capacity market cannot be a success without taking into account the technical characteristics of DR resources. First, to curtail demand, load resources need to be notified ahead of the time, and the notice time can range from 30 min to 2 h. Second, the number of times that the DRs may be called upon to provide a demand reduction varies with the specific market product and the capability of load resources. It can be a few times each year or unlimited. Third, some DRs may have more limited availability than generation resources. In some capacity markets, DRs can make themselves available only on a subset of hours in the summer, or only for a fixed number of calls while generation awarded in the capacity market has to be available throughout the year.

Markets with capacity obligations provide reasonably certain availability payments so that capacity markets have achieved the highest rates of DR participation up to this point, compared to energy and AS market. As the proportion of capacity supplied by load resources increases, the performance of DRs when called has to be examined. To procure capacity from DR providers that were not available year round could also adversely impact the grid reliability. Recent changes of the market rules ensure that DR providers respond to dispatch instructions when called and will be available in the year-around programs so that system operators can reliably deploy them in emergencies.

1.6.2 Demand Responses in North American Markets

The wholesale markets differ in their market rules and regulatory requirements, and therefore result in different participation levels of DRs in these markets. The following describes the current status of DR programs in four wholesale markets in North America.

1.6.2.1 ERCOT

The Electric Reliability Council of Texas (ERCOT) operates day-ahead and real-time energy markets with a current price cap of $9000/MWh. ERCOT does not have a resource adequacy requirement and thus does not operate a capacity market. The ERCOT market allows several types of demand response to participate in both energy and AS market. Through the load acting as a resource (LaaR) program, load resources provide a considerable amount of responsive reserve (1150–1790 MW). ERCOT has also recently developed a market rule so that CLRs can submit demand-side bids in the energy market dispatch although its participation is very low. CLRs participating in the real-time market are required to follow a 5-min dispatch signal sent from the system operator. The compensation for the load reduction in energy market is equivalent to the wholesale price minus an estimate of the avoided retail cost, which is termed "LMP minus G." Although ERCOT is an energy-only market, its Emergency Reserve Service (ERS) program has some similar characteristics with DR programs in capacity market, i.e., it procures DR to respond to system emergencies through an auction with an expenditure cap and pays DR providers an availability payment. ERS program distinguishes between loads with different willingness to be curtailed. Thus, when future load shedding is called, it first curtails customers with a relatively low value of lost load (VOLL).

1.6.2.2 PJM Interconnection

The PJM interconnection began paying loads for curtailment during emergency conditions in 2000. Since then, the markets have evolved and load resources are eligible to participate in PJM's energy, ancillary service, and capacity market. Today, DR resources become a significant part of the PJM wholesale market.

PJM's capacity market ensures long-term grid reliability by procuring the power supply resources needed to meet predicted energy demand 3 years in the future.

Since capacity market participation is more rewarding to DR providers, DR participation in PJM's capacity market has grown significantly over time, with more than 10,000 MW of DRs cleared in the forward capacity auction each year. This typically represents 5% of the annual committed capacity and its value exceeds $500 million. Recent change in the market rule also requires DR resources to reduce load throughout the year, which makes DR resources more comparable to generation.

In PJM, DRs can participate in the energy market either as an economic DR resource or a load management (capacity commitment) DR resource. A retail customer can be both an economic resource and a load management resource.

Load resources that participate as economic DR at PJM may be dispatched on an economic basis through security-constrained economic dispatch (SCED), similar to other capacity resources based on their availability. Despite the possibility of payments at full LMP, only a small amount of load management is actually dispatched on an economic basis. This is attributed to the fact that while each load resource with capacity commitment has an energy offer price that can be used for dispatch, the vast majority of them offer near or at the energy offer price caps. As a result, virtually all the load resources are dispatched by PJM through administrative actions.

Through out of the market operations, PJM will dispatch load resources when expecting a shortage on reserves. In order to know the effect of load reductions, PJM requires curtailment service providers (CSPs) to provide an hourly estimate of their expected real-time energy load reductions by considering the most accurate and latest information on the status of their DR resources. This estimate is provided to PJM to help guide its operation decisions.

Currently, the amount of load resources, which is registered to participate in PJM's energy market, is 3495 MW for economic DR and 12,866 MW for load management resources with a capacity commitment, respectively. On high energy price days, less than 1000 MW of economic DR could be available and dispatched at maximum. This low participation is likely due to less revenue opportunities in energy market relative to the emergency program or capacity market. In 2014, the emergency program accounted for nearly 97% of payments to DR providers in PJM, compared to 2.5% for the energy market program and 0.7% for synchronized reserves.

DR is also allowed to provide day-ahead schedule reserves (DASR), synchronized reserves (SR), and regulation reserve in PJM. DR may provide up to 25% of the overall DASR and regulation amounts and up to 33% of the SR requirement. The participation limit on DR was developed based on concerns about the performance of DRs when deployed.

1.6.2.3 MISO

MISO incorporated Demand Response Resource (DRR) into energy market which started in 2005. When its AS market began in 2009, MISO split DRR into DRR-Type I and DRR-Type II, and implemented enhanced DRR rules in 2015 under FERC order 719. DRR-Type I represents interruptible load or behind meter generation that can follow on/off instruction with an offered target demand reduction. As it is not dispatchable, DRR-Type I is only eligible to provide energy and contingency reserve (either spinning reserve or supplemental reserve). DRR-Type II represents behind meter generators or dispatchable load that can follow dispatch instructions and it is eligible to provide energy, regulating reserve, spinning reserve, and supplemental reserve. There are about 20 resources of approximate 470 MW in total participating in energy and ancillary service market in MISO and a majority of them are DRR-Type I. Only one DRR-Type II resource from a dispatchable load provides energy, regulating reserve, and spinning reserve. When deployed, DRR-Type II performs better than most of other resources in the regulation market.

In MISO, two types of demand response programs have been developed for resource adequacy under emergency conditions: Load modifying resource (LMR) and Emergency demand response (EDR). Both of them can be cleared in the capacity market to meet planning reserve margin requirement.

LMR is a category of demand resources or behind meter generation that can meet resource adequacy requirements and must be available during emergency conditions for demand reduction. Over 5000 MW of LMR have been cleared in MISO without an energy offer. They are compensated through capacity market auction and will be penalized if not performing satisfactorily during emergency events. When called, they will be provided with no more than 12 h advance notice from MISO and need to maintain a target level of demand reduction for at least 4 consecutive hours. In addition, they should be capable of being interrupted at least the first 5 times during the summer season when called upon by MISO.

EDR was created to enable more demand resources to assist in the grid operations during North American Electric Reliability Corporation (NERC) Emergency Alert 2 or Alert 3 (EEA2 or EEA3) events. About 900 MW of load resources participate in EDR at MISO. However, EDR resources do not have obligation to curtail during emergencies and can change their offer and availability day-by-day. When MISO calls on EDR for deployment, it is based on economic merit order from the offers. EDR market participants are not compensated through capacity market. Instead, they will be paid the higher of the revenues resulting from hourly LMPs, or the EDR resource's production costs for the total period of reduction. EDR production costs are defined as the shutdown cost plus the lesser of the amount of hourly demand reduction or the hourly dispatch instruction, multiplied by the EDR curtailment price applicable to the period of actual demand reduction.

1.6.2.4 ISO New England (ISO-NE)

ISO-NE's wholesale power system can be categorized into three markets with various products: energy, capacity, and ancillary service. The vast majority of load resources that participate in ISO-NE's energy market also participate in its annual forward capacity market held 3 years in advance of each commitment period. DR resources are allowed to participate in the energy and forward capacity markets, provided that the resources meet certain minimum qualification requirements.

ISO-NE designed the nation's first forward capacity market. The forward capacity market required all resources seeking to provide capacity to qualify for and participate in an auction that would occur 3 years in advance of the required delivery date. Load resources can participate in capacity market through either Real Time Demand Response or Real Time Emergency Generation program. Real Time Demand Response refers to a reduction in energy usage at an end-use customer facility, while Real Time Emergency Generation refers to an on-site generator behind the customer meter that has environmental permits limiting its operation to "emergency" hours when the system operator calls upon in order to prevent the load shedding. Both types receive the monthly capacity payment and in return must respond within 30 min when dispatched by ISO-NE.

In ISO-NE's capacity market, load resources are allowed to participate only for a limited period of time. On-peak resources offer on their reduced electricity consumption during summer peak hours and winter peak hours, while seasonal-peak resources offer on their reduced electricity consumption during the summer months and during the winter months when the real-time system hourly load is higher than certain level.

In New England, nearly 2000 MW of customer demand have accepted an obligation to reduce demand during emergency situations in return for a monthly reserve payment that has averaged slightly more than $3.50/kW-month. An examination of past events also indicates that those customers were dispatched rarely. Real Time Emergency Generation resources were dispatched region-wide on only 2 days in August 2006 for a total of 13 h, and 3 days in 2010 for a total of 11 h. In actual events, the overall performance of demand response resources in the New England capacity market relative to their obligation is also satisfactory. For example, in the event of June 24, 2010, when a capacity deficiency problem happened, of the 669 MW of demand response called, 653 MW responded and reduced demand within 30 min.

1.7 Lessons Learned and Outlook

The future of DRs will be more promising than today as DR programs are being gradually expanded into the whole spectrum of utility practices from long-term planning to real-time operations. This section summarizes the important lessons learned from the past experiences and presents a path toward a future power grid where the DR participation is high.

1.7.1 Lessons Learned

DR is a change in the power consumption of a load resource to better match the demand for power with the supply. Until recently electric energy could not be easily stored, so that utilities or grid operators have traditionally matched demand and supply by changing the power of thermal power plants or taking generating units on or off line. There are limits to what can be achieved on the supply side due to either high cost, low efficiency, or lack of generation capacity in the scarcity conditions. DR seeks to adjust the demand for power instead of adjusting the supply. The field experiences have demonstrated both technical merits and cost effectiveness of DRs. The reasons why DRs have been a success around the world can be summarized as follows.

First, DR resources have varied capabilities and services that they can provide, just as generation resources do. Smart grid technologies have improved the ability of load resources to communicate with the grid operators and make decisions about how and when to adjust electrical power demand according to the grid balancing need. This emerging technology not only allows for a precise control of load demand, but also shifts demand response from an event-based demand response where the utility requests the shedding of load, towards a more 24/7-based demand response where the customer sees incentives for controlling load all the time. As the DR control capability is improved, load resources do not need to be exempted from legacy obligations and requirements imposed on the market participation originally developed for generators, and thus are able to compete with generators in a level playing field.

Second, the participations of customers in DR programs are largely influenced by economic incentives. As the economic incentives remain the driving force behind the DR enrollment, the DR participation is only expanded when the gains from participating in DR programs outweigh the costs for enabling such a participation. The 2007 study from the Brattle Group quantified the value of demand response, and it was estimated that less than 2% load curtailment in PJM's market would reduce energy market prices between \$8 and \$25 per megawatt hour, or 5–8% on average, depending on market conditions. To date, the emergency program and capacity market still present the most lucrative market opportunity to DRs with a steady and predictable revenue resource.

Third, removing the market barriers is another condition prerequisite to expand the DR participation. The removal of these barriers requires engaging the entities that created the barrier which may not be inclined to make needed changes. Such efforts introduce additional costs, both in time and financial resources, in implementing the changes. These changes may also need an approval from federal regulators before being made to operating practices. The markets with the most successful DR programs have coped with these barriers within their retail and wholesale market environments, and engaged market participants to settle on a solution which could most benefit the market as a whole. With these barriers abolished, the markets lower the difficulty and risk of generating sufficient revenue to offset the cost of the load resource's participation in DR programs so that the DR participation eventually increases.

Fourth, a successful DR program requires the engagement of customers when they are enrolling in such a program. It is hard to expect a large number of consumers to continue their DR participation for monetary incentives only without other concerns being appropriately addressed. Load customers could be reluctant to relinquish total control of their assets to utility companies. They may be also worried with the protection of the privacy of data which are collected by the aggregation entities or third-party to facilitate the enrollment in DR programs. The customers may drop out of the DR programs if they are called upon for their response too often. To address all of the issues, the customers need to be informed about the significance and opportunities of DRs. A utility can only do a good job of recruiting and retaining customers for its DR programs when these customers are well educated and fully engaged.

Lastly, the potentials of residential customer demand response have been exploited through some levels of aggregations. Using dedicated control systems, DR schemes are currently implemented with large and small commercial as well as residential customers. An ensemble of small loads has the ability to provide high quality grid services, but avoids the burdens on each load who will not likely participate in DRs individually. An aggregating entity groups a suitable portfolio of assets. On the other hand, it delivers the obligated amount of grid services at the right times. Enabling hardware and software solutions and the commercial aggregator make appliances such as refrigerators, water heaters, HVAC systems, and clothes dryers to acquire new technical ability so that they can participate in the market easily and actively.

1.7.2 Changes for Future Grid to Enable High DR Participations

A future power grid definitely needs to bring about more changes to its structure and operation platform to enable a high participation level of DRs while maintaining its reliability, security, and efficiency.

1.7.2.1 Market Rules

Given the criticality of market rules to future development of DRs, more work and efforts to improve market designs need to be committed.

The amount of DRs entering the whole market is largely driven by how the market rules are created, which play a critical role in influencing the long-term viability of DR. Enhanced or new market designs can open up the market to more demand response resources while meeting future system reliability needs. For example, these changes could be to modify existing capacity programs or to develop separate programs for specific demand response resources that supply regulation or balancing services.

In an open wholesale market, market participants can buy or sell energy and service products in a fair, efficient, and non-discriminatory manner. Markets only work best when there is vigorous and voluntary participation by both buyers and sellers. To this end, demand responses can enter a market either in the demand side or in the supply side. The latter happens when the market provides load resources opportunities to participate on a comparable basis to supply side resources. A good market design not only does not discriminate DRs, but also establishes price signals for demand resources to make flexible energy use decisions and provides full cost recovery of the flexibility offered by demand resources.

However, a good market design is not guaranteed unless it has addressed the fundamental questions on the value of DRs, pricing mechanism, and rules governing contract performance and pricing elements. Each market differs in the generation mix and load profile, and accordingly, the need for new market product is also different. To evaluate the feasibility of including DR in an existing or new market product, a market analysis should be conducted to determine the value, appropriate timing, and quantity of DR, while giving consideration to costs to implement, benefits, and impacts. Design decisions need to be made on how the price is determined and how the curtailed load is compensated (availability payments vs. utilization payments). To accommodate the technical characteristics of demand response, the technical performance requirements for DR could deviate from those for generators, e.g., seasonal variations and different response time. To ensure the performance assurance, new mechanisms for the measurement and verification of load curtailment relative to the expected demand response are also needed.

1.7.2.2 Price Formulation

The activation of DRs may result in a depressed real-time price, and this undesirable consequence needs to be corrected by proposing an appropriate price formulation.

In North America, DRs are accounting for a large portion of capacity in emergency programs and some AS markets. In these programs, DR products may be dispatched outside of the market. In other words, emergency and AS DRs were not responding to the energy price when being curtailed, nor were they directly setting the price in times when they were curtailed. While deployment of these products helped the grid recover from contingency or emergency situations, activating these resources reduced load and could lead to reductions in energy prices during times of scarcity. Such price depression could be inappropriate due to conflict with the objectives of emergency actions in real time, and the actions of DRs could have the potential to endanger reliable operation. The problem is especially aggravated when the DR participation is high.

Prices in scarcity hours play a crucial role in sending efficient long-term economic signals to maintain adequate supply resources and to develop additional demand-response capability. Since current price is not efficient under emergency conditions, MISO has developed Extended Locational Marginal Pricing (ELMP), which partly addresses this issue by allowing block loaded emergency capacity resources to set price, but will not help when the emergency capacity resources are offered free or cheaper than the marginal resource dispatched prior to invoking emergency. ELMP was developed to more accurately reflect the cost of actions taken to meet demands as it allows block loaded fast-start resources (FSR), Emergency Demand Response resources, and offline FSR to set price. ELMP method does not change commitment and dispatch and is implemented as a pricing engine based upon existing day ahead and real-time economic dispatch software on single-interval basis. Future works are still ongoing to improve the performance of ELMP.

1.7.2.3 Coordination between DRs to Reduce "Paid Back" Energy

Cautions must be exercised when releasing load resources from their curtailment to avoid creating an artificial demand peak. It is a common assumption that a population of loads will behave in a predictable manner when deployed for demand response. However, this becomes invalid as the states of loads take homogeneous values if not coordinated well.

To curtail or postpone a large amount of DRs could make the load resources affected to lose diversity among their states or configurations after the services are restored or the curtailment is stopped. The load demand, by nature, has different start timing and cycle. However, in a case when decreasing energy consumption for a load with thermal inertia at a given time, a thermal mass or process must be deferred. For an HVAC system, this results in a warmer (or cooler) residence. For a hot water heater, this results in lower water temperature. However, these are energy limited resources. The HVAC system cannot defer operation indefinitely, so that it must eventually turn on. A dryer can operate at a lower temperature, but eventually the necessary energy must be consumed to dry the clothes. The amount of resource depends on technical as well as social limits and preference. Eventually the energy must be "paid back" (e.g., the house has to be cooled back down to normal air temperatures). When a large number of load resources are released simultaneously from their curtailment, they will start consuming energy at the same time, i.e., their cycles for drawing power from the grid are temporarily synchronized. With DR programs using automated technologies, a significant rebound can be experienced due to the synchronization of the starting time in consuming energy. One of these examples is depicted in Fig. 1.3. The blue and yellow traces show the demand profile as the baseline without curtailment and the actual load, respectively, and the energy price is shown in gray. For this particular event, due to "paid back" energy, the actual load spike immediately after the completion of curtailment was even higher than the load demand if demand response was not called upon. This emphasizes a need for coordinating the actions among load resources so that they can behave in a controllable and smooth way as designed.

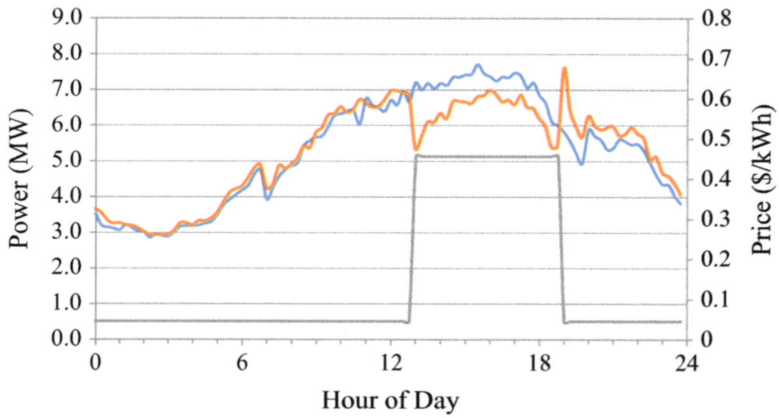

Fig. 1.3 Illustration of "paid back" effect of demand response

1.8 Summaries

Demand response is a reduction in load demand designed to respond to pricing signals or the system dispatch instruction. Since invented in the 1970s, DR programs are being used by electric system planners and operators as resource options for balancing supply and demand. Such programs can lower the cost of electricity in wholesale markets and, in turn, lead to lower retail rates. Methods of engaging customers in demand response efforts include offering time-based rates such as time-of-use pricing, critical peak pricing, variable peak pricing, real-time pricing, and critical peak rebates. It also includes direct load control programs which provide the ability for the utilities to interrupt load resources during periods of peak demand or emergencies in exchange for a financial incentive and lower electric bills.

The electric power industry considers DR programs as an increasingly valuable resource option whose capabilities and potential impacts are expanded by grid modernization efforts. DRs can be a more cost-effective alternative than adding generation capabilities to meet the peak demand. Advanced metering infrastructure expands the range of services provided by residential load consumers. Smart customer systems such as in-home displays or home-area-networks can make it easier for consumers to change their behavior and to adjust load demand so as to dynamically contribute to the balancing need.

Dispatchable or controllable load resources can participate in the wholesale market and compete with generation supply to provide energy, AS, or capacity. A high DR participation in capacity market or emergency programs has been seen in the USA due to considerable economic gain to load resources. As the market designs for load demand to enter into energy and AS market are being improved, the future of DRs is just unfolded.

DRs hold great potentials for future power grid, both from reliability and efficiency perspective. As the techniques for enabling and controlling DRs are evolving, more applications and business cases for DR programs will emerge. The future development of DRs necessitates a collaborative approach to engaging stakeholders and entities in removing market and regulatory barriers for a high DR participation.

References

1. Brown, T., Newell, S. A., Oates, D. L., & Spees, K. (2015). *International review of demand response mechanisms*. Boston: Brattle Group.
2. *Demand response strategy*. PJM interconnection, 2017.
3. Hurley, D, Peterson, P., & Whited, M. (2013). Demand response as a power system resource: Program designs, performance, and lessons learned in the United States. In *Synapse energy economics*.
4. Wang, C., Chatterjee, D., Li, J., & Robinson M. (2017) Price formation with evolving resource mix. *Mid-Continent ISO*.
5. Du, P., & Lu, N. (2011). Appliance commitment for household load scheduling. *IEEE Transactions on Smart Grid, 2*(2), 411–419.
6. Li, W., Du, P., & Lu, N. (2018). Design of a new primary frequency control market for hosting frequency response reserve offers from both generators and loads. *IEEE Transactions on Smart Grid, 9*(5), 4883–4892.
7. Chen, Y. (2016). Demand response resources in MISO and other RTOs. *IEEE PES General Meeting*.
8. Rahimi, F., & Ipakchi, A. (2010). Demand response as a market resource under the smart grid paradigm. *IEEE Transactions on Smart Grid, 1*, 82–88.
9. Callaway, D. S., & Hiskens, I. A. (2011). Achieving controllability of electric loads. *Proceedings of the IEEE, 99*(1), 184–199.
10. Conejo, A. J., Morales, J. M., & Baringo, L. (2010). Real-time demand response model. *IEEE Transactions on Smart Grid, 1*(3), 236–242.
11. Chassin, D. P. (2017). *Multi-scale transactive control in interconnected bulk power systems under high renewable energy supply and high demand response scenarios*.
12. Cui, W., Ding, Y., Hui, H., Lin, Z., Du, P., Song, Y., & Shao, C. (2018). Evaluation and sequential dispatch of operating reserve provided by air conditioners considering lead–lag rebound effect. *IEEE Transactions on Power Systems, 33*(6), 6935–6950.
13. Liu, C., & Du, P. (2018). Participation of load resources in day-ahead market to provide primary-frequency response reserve. *IEEE Transactions on Power Systems, 33*(5), 5041–5051.
14. Xu, Q., Ding, Y., Yan, Q., Zheng, A., & Du, P. (2017). Day-ahead load peak shedding/shifting scheme based on potential load values utilization: Theory and practice of policy-driven demand response in China. *IEEE Access, 5*, 22892–22901.
15. Du, P., Baldick, R., & Tuohy, A. (2017). *Integration of large-scale renewable energy into bulk power systems*. Berlin: Springer International Publishing.
16. Pinson, P., & Madsen, H. (2014). Benefits and challenges of electrical demand response: A critical review. *Renewable and Sustainable Energy Reviews, 39*, 686–699.

Chapter 2
Modeling Demand Response Resources

2.1 Thermal Characteristics of the TCL

Because the operation principle of all TCLs is similar, we will use the modeling of an HVAC unit as an example to describe the basic modeling principles.

To model the electricity consumption of an HVAC unit, it is critical to model the unit's heat transfer process considering the change of ambient temperatures. Figure 2.1 depicts the thermal behavior of a residential house in winter over time. In winter, the HVAC unit operates in the heating mode. As observed, the room temperature varies within a deadband, T^{DB}. The rising curve indicates that the space heater unit is "on," and the falling curves represent the cooling down periods, when the heater is "off." As the space heater cycles, the room temperature rises and falls accordingly. T^{+} and T^{-} are the upper and lower limits for a given thermostat setpoint, T_{set}, such that $T^{DB} = T^{+} - T^{-}$ and $T^{+} = T_{set} + 0.5T^{DB}$. The unit has a cycling time of τ., with an "on" period of τ_{on} and an "off" period of τ_{off}. The asymptotic equilibrium temperature, T_e, is usually far below the thermostatic setpoint by design. This makes the temperature change rate within T^{DB} almost linear, as shown in Fig. 2.1. This important characteristic warrants the use of the first order differential equation for modeling the temperature variations instead of seeking higher order models.

Changing T_{set} allows one to regulate the energy consumption of the TCL in a given period *T*. For example, if T_{set} is increased from 75 to 76 °F at time *t*, the "on" cycle will be longer and the "off" cycle will be shorter, so that the overall energy consumption of the space heater will be increased for that day. However, at time *t*, if the space heater is "off" and the room temperature is above 76 °F, after the setpoint is increased to 76 °F, the space heater will not be immediately turned on. Therefore, raising T_{set} may not immediately result in a power consumption increase. Similarly, decreasing T_{set} may not immediately lead to a power consumption decrease.

Turning on a TCL when it is in the "off" cycle can instantaneously increase the power consumption at time *t* and turning off a TCL when it is in the "on" cycle can

P. Du et al., *Demand Response in Smart Grids*,
https://doi.org/10.1007/978-3-030-19769-8_2

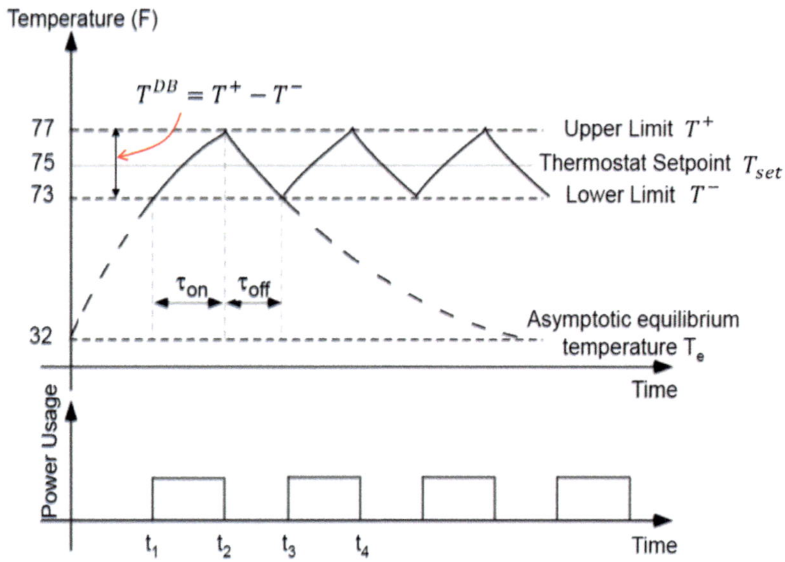

Fig. 2.1 An illustration of the room temperature profile and the HVAC on/off curve of a residential HVAC system

instantaneously decrease the power consumption at time t. However, if T_{set} remains unchanged, the overall energy consumption in T may not change significantly because the overall energy required to heat up the room remains the same unless there is a huge ambient temperature change. Therefore, when using TCLs as resources for DR, a decision made at time t may impact their operating status in the subsequent time steps. This is the main reason why the thermal behaviors of the TCL, i.e., the cycling characteristics for the rest of the control period, need to be modeled.

In general, there are two TCL modeling approaches: the first-principle-based approach and the measurement-based approach. Because each approach has its distinct advantages and disadvantages for predicting TCL cycling characteristics, it is important for the modelers to understand the assumptions and simplifications made to derive the modeling parameters, so that they can select a suitable approach for their specific applications.

2.2 First-Principle-Based Approach

The first-principle-based or physics-based approach takes into account the thermal characteristics of the TCL such as the thermal resistance of the building materials, the thermal capacitance for a thermal mass to store heat, and the heat injection. It usually requires a bottom-up approach to build a thermal model that reflects the

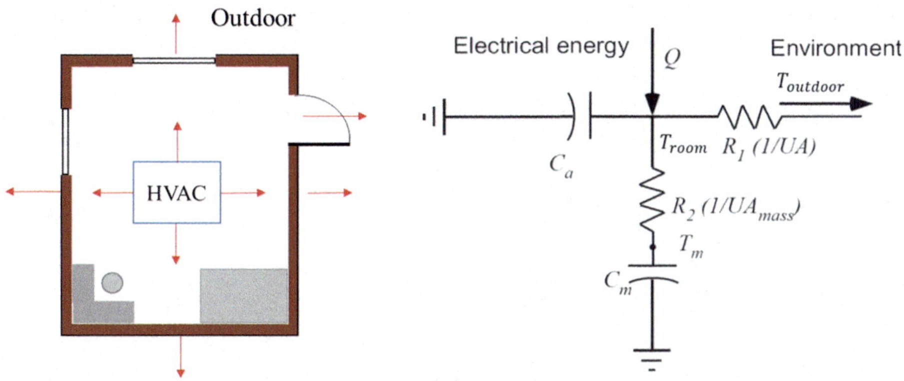

Fig. 2.2 Equivalent electric circuit of a residential HVAC system

thermal dynamics and converts the cycling characteristics of TCL into the ON/OFF cycles of the TCL in order to calculate the electricity consumed. The most commonly used first-principle-based method is the equivalent thermal parameter (ETP) model. An ETP model of an HVAC unit is depicted in Fig. 2.2.

The furniture and air inside a room are thermal mass and can store heat, making them equivalent to capacitors in an electric circuit. Heat can escape from the house through walls and windows, and the rate of the heat dissipation is determined by their standby heat loss coefficients, making them resemble resistors in an electric circuit. Electricity is converted to heat when the HVAC is in its heating mode, so Q is positive and represents the heat injected by the heater. When the HVAC is in its cooling mode, Q is negative and represents the heat extracted by the compressor.

If we use the state space model to represent the thermal dynamic process of the HVAC unit, we have

$$\begin{aligned} \dot{x} &= Ax + Bu \\ y &= Cx + Du \end{aligned} \tag{2.1}$$

where

$$\dot{x} = \begin{bmatrix} \dot{T}_i \\ \dot{T}_m \end{bmatrix} \quad x = \begin{bmatrix} T_i \\ T_m \end{bmatrix} \quad u = \begin{bmatrix} T_{outdoor} \\ Q \end{bmatrix}$$

$$A = \begin{bmatrix} -\left(\frac{1}{R_2 C_a} + \frac{1}{R_1 C_a}\right) & \frac{1}{R_2 C_a} \\ \frac{1}{R_2 C_m} & -\frac{1}{R_2 C_m} \end{bmatrix} B = \begin{bmatrix} \frac{1}{R_1 C_a} & \frac{1}{C_a} \\ 0 & 0 \end{bmatrix} C = \begin{bmatrix} 1 & 0 \\ 0 & 1 \end{bmatrix} D = \begin{bmatrix} 0 & 0 \\ 0 & 0 \end{bmatrix}$$

C_a	air heat capacity (Btu/°F or J/°C)
C_m	mass heat capacity (Btu/°F or J/°C)
Q	heat rate for the HVAC unit (Btu/h or W)
UA	standby heat loss coefficient of the wall (Btu/°F·h or W/°C)
UA_{mass}	standby heat loss coefficient of the thermal mass (Btu/°F·h or W/°C)
R_1	1/UA (°F·h/Btu or °C/W)
R_2	1/UA$_{\text{mass}}$ (°F·h/Btu or °C/W)
$T_{outdoor}$	ambient temperature (°F or °C)
T_{room}	air temperature inside the house (°F or °C)
T_m	mass temperature inside the house (°F or °C)

We call this model the state-space RCQ (SS-RCQ) model. The typical ranges of parameters used in the SS-RCQ model are listed in Table 2.1. By randomizing the parameters within a given range, the modeler can generate different sets of parameters for each household in order to model the diversity in the cycling characteristics of the HVAC units.

Example 2.1 The model parameters of an HVAC unit are given in Table 2.2. Assume that the outdoor temperature $T_{outdoor}$ is hold constant for a 1-h period.

1. Calculate the ABCD matrices of the SS-RCQ model.
2. Plot the room temperature profiles when $T_{outdoor}$ is hold at 0 °C, 5 °C, and 10 °C, respectively.

Solution The MATLAB code example 2_1.m is used to calculate the HVAC parameters and plot Fig. 2.3.

Although the SS-RCQ model is developed for the HVAC unit, the same modeling principle can be used to model refrigerators and electric water heaters. There are a few additional considerations when modeling water heaters. First, the consumption of hot water needs to be considered. For example, taking a hot tub requires a large amount of hot water but washing hands or fruits require a small amount of hot water draw. The former happens less often but will cause the water heater to turn on immediately; the later happens more frequent but may only shorten the "off" cycle by minutes. Second, because there is a strong correlation between the water heater load and washing loads (e.g., washers and dishwashers), the modeling of those loads need to be coordinated. Interested readers can refer to [2, 3] for more details.

The first-principle-based method is a rigorous analytical modeling method with all parameters bearing physical meanings. Thus, sensitivity analysis can be conducted to address the impacts of each physical parameter (e.g., R, C, and Q) or control setting (e.g., T^{DB} and T_{set}) on the cycling characteristics τ_{on} and τ_{off}.

If multiple HVAC units are used in one house or for a commercial building, using one set of RCQ values or using a first order ETP model may not be sufficient. However, to model demand response programs, often times, only a lump-model is required to calculate the electricity consumption. Therefore, in those cases, the RCQ model remains a good option.

Table 2.1 Range and Typical Values of the Key Input Parameters of an HVAC unit

Parameter	Area (f^2 or m^2)	UA (Btu/·°F or W/°C)	Thermal Mass (Btu/F or W/°C)	Deadband (°F or °C)	Rated Cooling Capacity (Btu/h. or W)	Indoor Temperature Setpoint (°F or °C)
Range	1500–3500 (139–325)	(0.15–0.35) × area (0.8–0.18)	(1–5) × area (0.53–2.64)	±0.5–2.0 (0.28–1.1)	(18–36) × area (5.3–15.8)	75–76 (23.9–24.4)
Typical value	2000 (186)	400 (211)	4000 (2110)	1.0 (0.56)	48,000 (14,064)	75 (23.9)

Table 2.2 An example of the HVAC model parameters used in simulations

Parameter	Area (m^2)	UA (W/°C)	Thermal Mass (W/°C)	Deadband (°C)	Rated Cooling Capacity (W)	Indoor Temperature Setpoint (°C)
HVAC	228	111	3924	1.0	5200	20

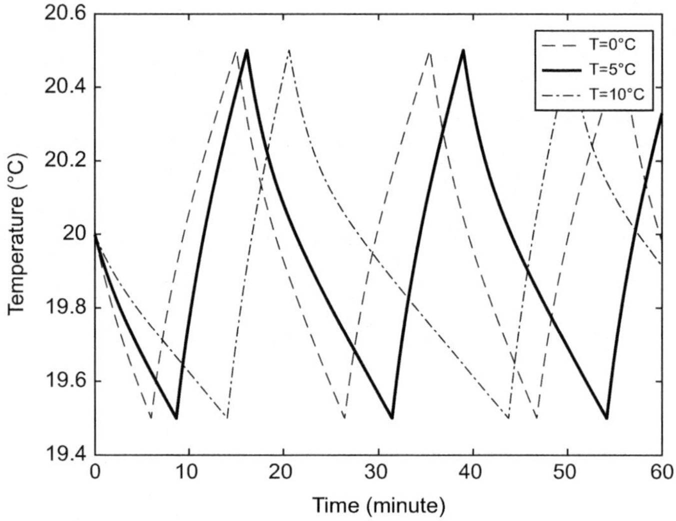

Fig. 2.3 Room temperature profiles when $T_{outdoor} = 0$ °C, 5 °C*and* 10 °C. $T_{set} = 20$°C and $T^{DB} = 1$ °C)

Generally speaking, the first-principle-based method usually demands detailed understanding of the thermal characteristics of the TCL. For example, for the HVAC unit, one needs to know the thickness of the wall, the facing of the house, the number of windows and doors, and the heat capacity of the furniture and air, etc. Therefore, the first-principle-based method is often used to develop a generic model for studying the aggregated load profile of a group of thermostatically controlled units before and after the implementation of a demand response algorithm.

2.3 Measurement-Based Approach

Because of variations in user behaviors and in the physical characteristics of the TCL device, the modeling errors of the first-principle-based method are usually too large to be used for real-time DR control needs. For example, customers' preference may differ from house to house regarding how the thermostat is set and adjusted, how often the windows and doors are closed, and when and how long the cooking activity lasts, etc. Even though the first-principle-based method starts with an accurate set of initial conditions and captures all the physical parameters of the house, the forecasted cycling characteristics will no longer be accurate due to the influence of the occupants' activities and ambient conditions (e.g., solar radiation, humidity, shades, etc.).

A measurement-based approach requires a modeler to construct a model from data. The widespread of advanced metering infrastructure (AMI) and sensor networks enables access to large amounts of historical data. Therefore, measurement-based approaches are usually preferred when used to forecast the TCL cycling characteristics in real-time DR control applications. There are two key modeling considerations. *First*, a lower order model with less parameters is preferred. *Second*, a periodically error correction process is often necessary to correct the modeling error cumulated over time.

In the subsequence sections, we will derive the ETP model used in the measurement-based approach and introduce the error correction process.

2.3.1 Simplified RCQ Model

The general approach of the measurement-based method is to start with a generic TCL model derived from the first-principle-based method that can be used for parameter identification. If we lump the thermal capacity of the air and all thermal mass and ignore the heat loss of the thermal mass, we can simplify the ETP model formulated by (2.1) and depicted in Fig. 2.2 to a one capacitor and one resistor ETP model, as shown in Fig. 2.4.

Then, the solution of the SS-RCQ model (2.1) in time domain of the simplified RCQ model is

$$u_{HVAC}^{t+1} = \begin{cases} 1 & \text{when} \quad T_{room}^{t+1} < T^{-} \\ u_{HVAC}^{t+1} & \text{when} \quad T^{-} \leq T_{room}^{t+1} \leq T^{+} \\ 0 & \text{when} \quad T_{room}^{t+1} > T^{+} \end{cases} \tag{2.2}$$

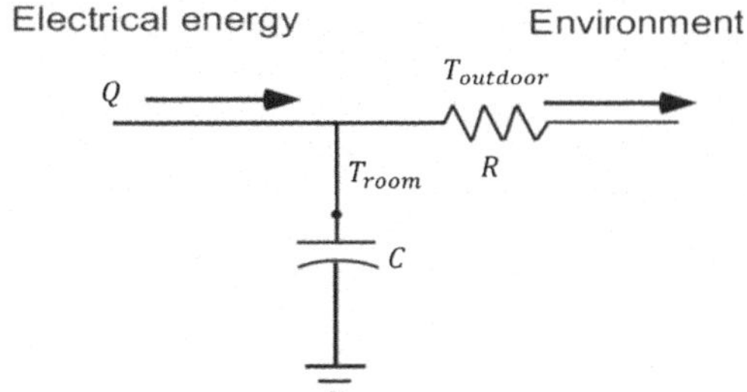

Fig. 2.4 Equivalent electric circuit of a residential HVAC system

$$T_{room}^{t+1} = \begin{cases} T_{outdoor}^{t+1} - \left(T_{outdoor}^{t+1} - T_{room}^{t}\right)e^{-\frac{\Delta t}{RC}} & \text{when the HVAC is “off”} \quad \left(u_{HVAC}^{t+1} = 0\right) \\ T_{outdoor}^{t+1} + QR - \left(T_{outdoor}^{t+1} + QR - T_{room}^{t}\right)e^{-\frac{\Delta t}{RC}} & \text{when the HVAC is “on”} \quad \left(u_{HVAC}^{t+1} = 1\right) \end{cases} \tag{2.3}$$

where.

C	equivalent heat capacity (J/°C)
u_{HVAC}^{t+1}	Status of the HVAC unit (“1” is “on”; “0” is “off”)
R	Equivalent thermal resistance (°C/W)
Q	Equivalent heat rate (W)
t	Time (second)
Δt	Time step (second).

A valid set of measurements include the turn-on time τ_{on}, the turn-off time τ_{off}, and its corresponding ambient temperatures $T_{outdoor}$ at a given setpoint T_{set} and T^{DB}. Parameters to be calculated are R, C, and Q, which can be calculated from

$$T^{-} = T_{outdoor} + QR - \left(T_{outdoor} + QR - T^{+}\right)e^{\frac{-\tau_{on}}{RC}} \tag{2.4}$$

$$T^{+} = T_{outdoor} - \left(T_{outdoor} - T^{-}\right)e^{\frac{-\tau_{off}}{RC}} \tag{2.5}$$

If we have N sets of measurements, we will obtain N sets of RCQ values. Then, the linear least squares fitting technique can be used to calculate an optimal set of values, $\{R^*, C^*, Q^*\}$, that provides the best fit to all N sets of measurements. Then, using $\{R^*, C^*, Q^*\}$, (2.2) and (2.3) can be used to predict the HVAC behaviors.

The advantage of using the first-order model as described by (2.4) and (2.5) is its simplicity. Because the room temperature variation is controlled within a very narrow temperature band (e.g., 1 or 2 °C), the temperature changing process is pretty linear, as shown in Fig. 2.3. Therefore, using high-order models is, in general, not necessary.

The main disadvantage of the measurement based model is that τ_{on} and τ_{off} may change drastically when $T_{outdoor}$ changes. In addition, because there are solar radiations and more customer activities at day-times, the day-time and night-time RCQ values are not the same even though they are derived at the same $T_{outdoor}$. For example, the RCQ values derived for $T_{outdoor}$ in a range between 5 and 10 °F at the nighttime may result in a longer τ_{on} and a shorter τ_{off} when used in the daytime.

A straightforward approach to compensate for this modeling error is to derive a set of $\{R^*, C^*, Q^*\}$ for each $T_{outdoor}$ range. Another option is to use the measurement data to correct the modeling errors in real-time. The third option is to decouple the modeling of the ON/OFF cycles. In this book, we will illustrate the first two options.

2.3.2 Linearized ETP Model

Because the asymptotic equilibrium temperatures are generally far beyond the room temperature setpoint for appropriately sized equipment, the exponential rising curve and falling curve are almost linear between the upper limit and the lower limit, as shown in Fig. 2.3. Therefore, a further simplification of (2.2) and (2.3) is to use linearized turn-on and turn-off curves to forecast the TCA temperature. The mathematical formulation of the linearized ETP model is

$$T_{room}^{t+1} = \begin{cases} T_{room}^{t} + \dfrac{T^{DB}}{\tau_{on}} \Delta t \\ T_{room}^{t} - \dfrac{T^{DB}}{\tau_{off}} \Delta t \end{cases} \tag{2.6}$$

As shown in Fig. 2.5, the turn-on time τ_{on} and the turn-off time τ_{off} of an HVAC in its cooling mode are functions of $T_{outdoor}$. After curve fitting, we can obtain two sets of functions that reflect the impact of the outdoor temperature on τ_{on} and τ_{off}.

$$\tau_{on} = b_1 {T_o}^2 + b_2 T + b_3 \tag{2.7}$$

$$\tau_{off} = a_1 {T_o}^4 + a_2 {T_o}^3 + a_3 {T_o}^2 + a_4 T_o + a_5 \tag{2.8}$$

where a_i and b_i are curve fitting parameters.

After combining (2.6)–(2.8), we have

$$T_{room}^{t+1} = \begin{cases} T_{room}^{t} + \dfrac{T^{DB}}{\tau_{on}} \Delta t & \tau_{on} = b_1 {T_o}^2 + b_2 T + b_3 \\ T_{room}^{t} - \dfrac{T^{DB}}{\tau_{off}} \Delta t & \tau_{off} = a_1 {T_o}^4 + a_2 {T_o}^3 + a_3 {T_o}^2 + a_4 T_o + a_5 \end{cases} \tag{2.9}$$

The advantage of the linearized ETP model is that it is very easy to derive based on measurement data compared with the exponential model and it can compensate the modeling error caused by the outdoor temperature changes. The disadvantage is that it requires more parameters to calculate variations in cycling durations caused by ambient temperature changes.

Example 2.2 Using the model parameters of the HVAC unit given in Table 2.2, calculate the ON/OFF time τ_{on} and τ_{off} when $T_{outdoor}$ is −10 °C, −5 °C, 0 °C, 5 °C, and 10 °C, respectively. Assume that the outdoor temperature $T_{outdoor}$ is hold constant for a 1-h period. Apply polynomial curve fitting to calculate the third order polynomial fit parameters for the τ_{on} and τ_{off} with respect to $T_{outdoor}$.

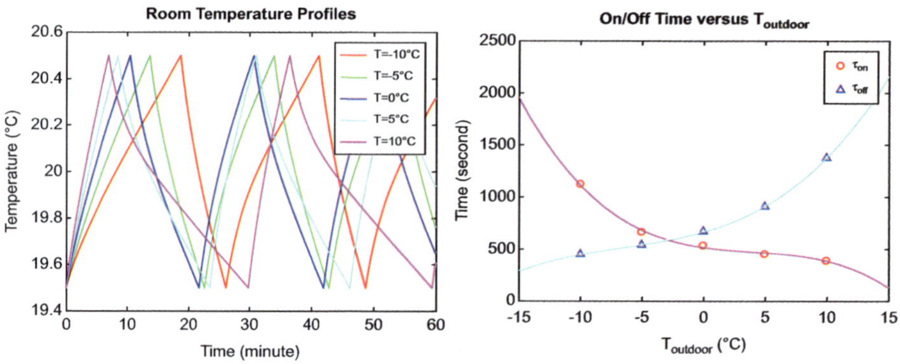

Fig. 2.5 Room temperature profiles and on and off durations of an HVAC unit in its heating mode when outdoor temperatures are at −10 °C, −5 °C, 0 °C, 5 °C and 10 °C, respectively

Table 2.3 ON/OFF times of the HVAC at different outdoor temperature

$T_{outdoor}$(°C)	−10	−5	0	5	10
τ_{on}(s)	1123	672	539	451	389
τ_{off}(s)	454	543	680	913	1385

Solution:

The room temperature changes are shown in Fig. 2.5 and the corresponding on/off cycles are shown in Table 2.3.

Solution:

The MATLAB code example 2_2.m is used to produce Fig. 2.5.

Observations:

In Fig. 2.5, the figure on the left depicts the temperature profiles predicted by the SS-RCQ model at different outdoor temperature; the figure on the right illustrates how the turn-on time τ_{on} the turn-off time τ_{off} of an HVAC will change with respect to the outdoor temperature changes. Although outdoor temperature changes within an hour is usually within 3 °C, if constant τ_{on} and τ_{off} are used without considering the outdoor temperature changes, the modeling error will soon become too large for the model to be used for predicting the HVAC cycling characteristics.

2.4 Error Correction by Measurement

Because the thermal process of a building is highly nonlinear and heavily influenced by occupants' random behaviors, large modeling errors will inevitably occur and accumulated over time, making modeled HVAC behaviors quickly deviate from the actual. Therefore, if a TCL model is used for predicting the behavior of an HVAC unit in real-time DR control, it is critical to correct the modeling errors periodically using measurements.

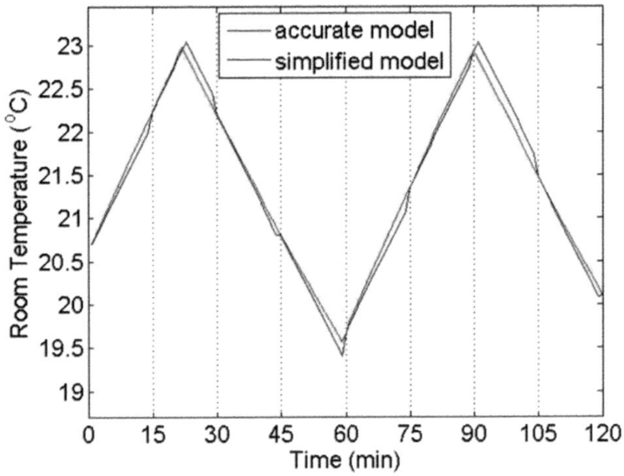

Fig. 2.6 Space heating unit behavior modeled by the simplified model and tuned by measurements (R, C, and Q are set to 0.1208 °C/W, 3599.3 J/°C, and 400 W, respectively)

To illustrate the error correction process, let's use the SS-RCQ model to produce the "actual" room temperature and use the linearized ETP model as the forecaster to produce the "forecasted" room temperature. Figure 2.6 shows how the "forecasted" room temperatures (the blue line) are tuned by the measured "actual" room temperature (the red line) every 15 min. As shown in the figure, every 15 min, a measurement data is received and used as the initial condition to perform the calculation of the room temperatures for the next 15 min interval. The modeling error accumulates for 15 min; then it is cleared by the measurement data.

In general, the shorter the error correction interval is, the less the modeling error will be. However, as more frequent communication between the temperature sensors and the forecaster requires more system resources to accomplish, a tradeoff needs to be made based on how much tolerances the customers can withstand when the room temperature is higher or lower than the preferred temperature limits.

In Fig. 2.6, if a DR controller switches on/off the HVAC unit based on the forecasted room temperature instead of the actual room temperature, the heater ON/OFF time can be longer or shorter than needed (i.e., cycling faster or slower). As a result, the room temperature may rise above or drop below the desired temperature band. If we update the room temperature forecast every 30 min instead of 15 min, the cumulated errors may cause the room temperature to raise 0.1 °C above 23 °C for one cycle. If the modeling errors are accumulated for a few more cycles, the room temperature deviations can be significant. In Example 2.3, we illustrate in more detail how the room temperature profile changes with respect to the error correction process.

Example 2.3 Model parameters and initial conditions of an HVAC unit are given in Table 2.2. Please calculate the ON/OFF time, τ_{on} and τ_{off}, for the following five cases assuming that $T_{outdoor}$ is increasing linearly from 0 to 5 °C in 3 h. Let's assume that the SS-RCQ model calculated room temperature at time t is the "actual" room temperature. Use the linearized ETP model to produce the "forecasted" room temperature. To simplify the calculation, use the ON/OFF time calculated when the outdoor temperature is 0 °C to calculate the temperature increase and decrease rate (i.e., τ_{on} is 538 s and τ_{off} is 680 s).

1. Assume that the HVAC unit is controlled by the local thermostat using the actual room temperature to determine the on/off of the HVAC unit.
   - Plot the "actual" room temperature profile produced by the SS-RCQ model. Plot the "forecasted" room temperature profile produced by the linearized ETP model using the same initial condition on the same figure, as shown in Fig. 2.7a.
2. Assume that the HVAC unit is controlled by a remote controller and the local thermostat is disabled. The remote controller will turn the HVAC unit on/off based on the "forecasted" room temperature.
   - No error correction case. Plot the "forecasted" and the "actual" room temperature profiles, as shown in Fig. 2.7b.
   - Let the errors be corrected every 15 min. Plot the "forecasted" and the "actual" room temperature profiles, as shown in Fig. 2.7c.
   - Let the errors are corrected every 30 min. Plot the "forecasted" and the "actual" room temperature profiles, as shown in Fig. 2.7d.
   - Let the errors are corrected every 60 min. Plot the "forecasted" and the "actual" room temperature profiles, as shown in Fig. 2.7e.

Solution:
The MATLAB code, Example 2_3.m, is used to produce Fig. 2.7. Note that the initial conditions are the same as in Example 2.1.

Observations:
The simulation results are shown in Fig. 2.7a–e.

As shown in Fig. 2.1a, as the outdoor temperature is increasing linearly from 0 to 5 °C in 3 h, using the linearized ETP model parameters obtained at 0 °C will result in a shorter cycling time. This modeling error will accumulate over time.

As shown in Fig. 2.1b, if we control the ON/OFF of the HVAC unit based on the room temperature forecasted by the linearized ETP model, the room temperature at the end of the 3-h period will exceed the upper temperature limit 20.5 °C.

As shown in Fig. 2.1c–e, if we use the measured room temperature to correct the modeling error every 15, 30, or 60 min, the level of room temperature excursions from the upper temperature limit will be reduced compared with the non-correction case shown in Fig. 2.1b. It can be seen that the more frequent the update is, the smaller the temperature violations are.

Interested readers are encouraged to read the error correction process introduced in [1, 4] provided in the "References" at the end of this chapter.

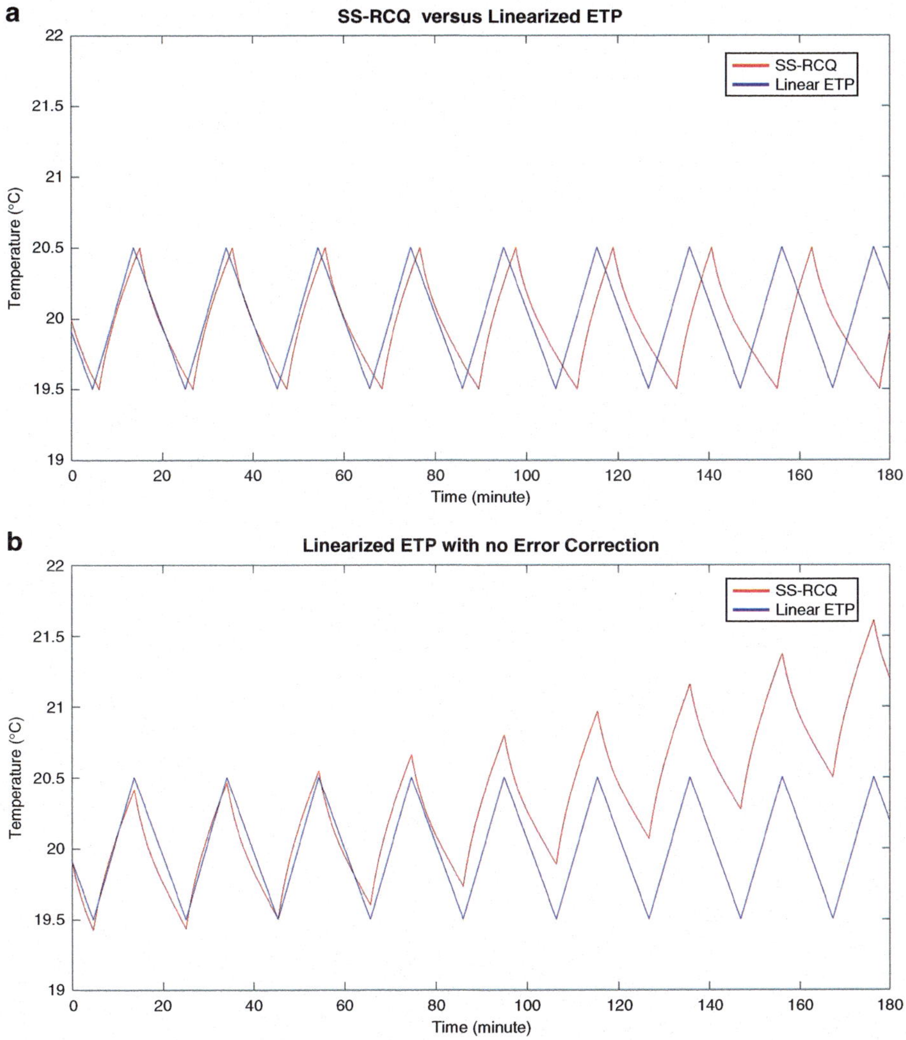

Fig. 2.7 (**a**) Room temperature profiles modeled by the linearized ETP model and the SS-RCQ model. (**b**) Room temperature profiles predicted by the linearized ETP model and the accurate room temperature simulated by the SS-RCQ model. (**c**) Room temperature profile predicted by the linearized ETP model and corrected by the SS-RCQ model every 15 min versus the accurate room temperature profile simulated by the SS-RCQ model. (**d**) Room temperature profile predicted by the linearized ETP model and corrected by the SS-RCQ model every 30 min versus the accurate room temperature profile simulated by the SS-RCQ model. (**e**) Room temperature profile predicted by the linearized ETP model and corrected by the SS-RCQ model every 60 min versus the accurate room temperature profile simulated by the SS-RCQ model

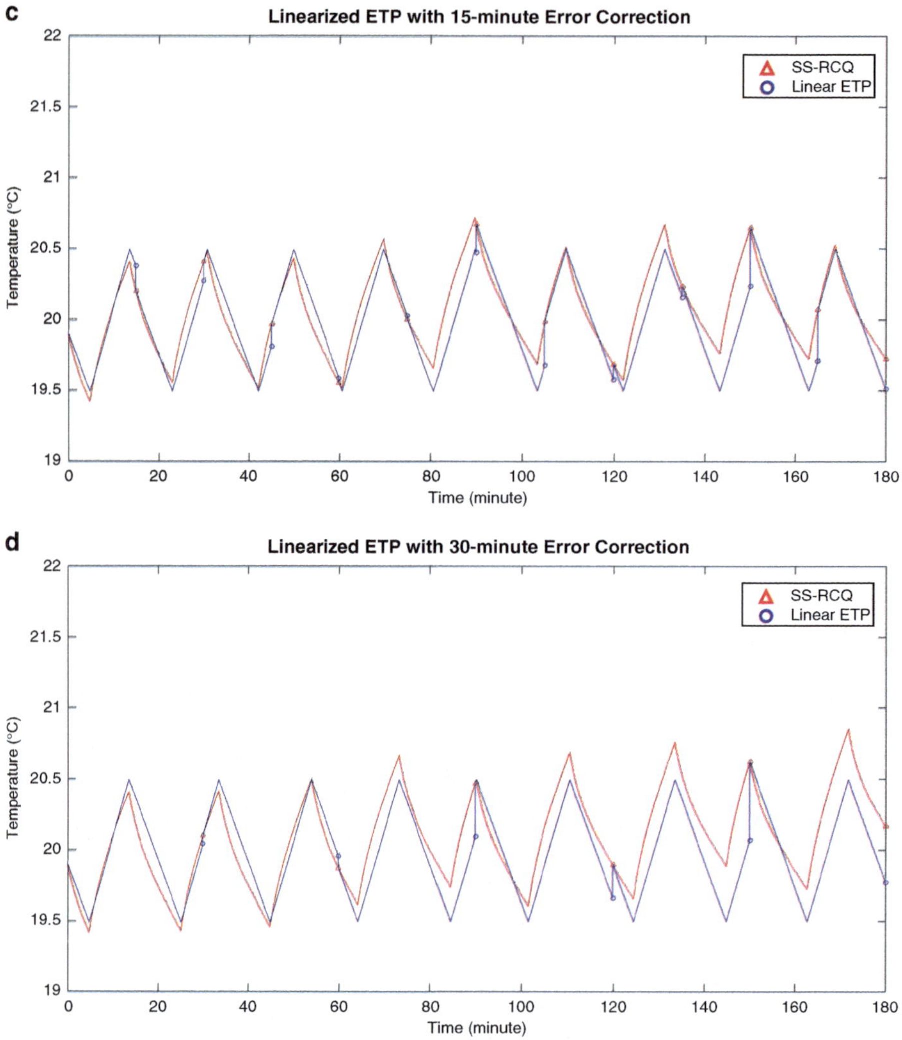

Fig. 2.7 (continued)

2.5 Parameter Randomization

In large-scale simulations, the RCQ parameter of an HVAC unit needs to be randomly selected when a heterogeneous group of HVAC units are modeled. There are two ways to randomize the RCQ parameters.

The first approach is to randomly select the RCQ values within given ranges, $[R^{lowlimit}, R^{highlimit}]$, $[C^{lowlimit}, C^{highlimit}]$, and $[Q^{lowlimit}, Q^{highlimit}]$, respectively. This approach is the most commonly used approach when modeling a DR group

e

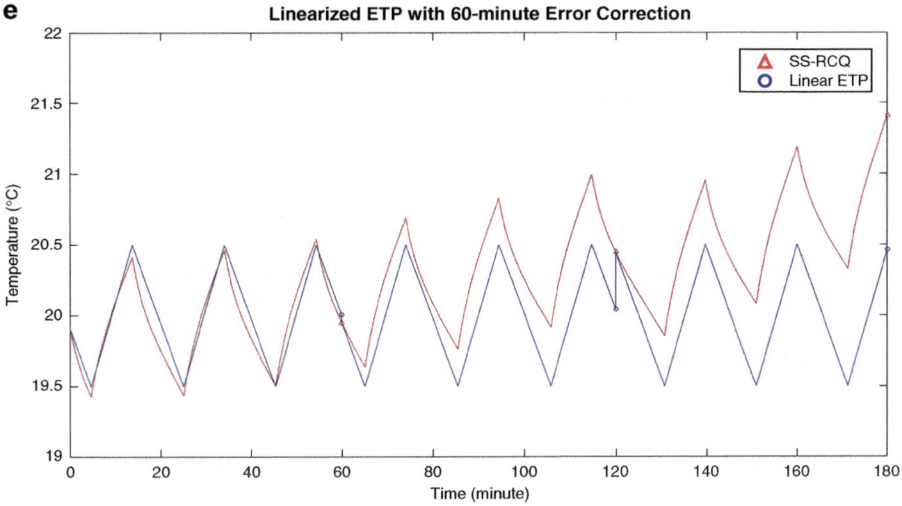

Fig. 2.7 (continued)

consisting of hundreds or thousands of HVAC units. The most commonly used PDFs for selecting the RCQ values within the given ranges are the normal or the uniform distributions.

In the second approach, we assume that the ON/OFF time τ_{on} and τ_{off} of a group of HVAC units follow a PDF within a given range $\left[\tau_{on}^{lowlimit}, \tau_{on}^{highlimit}\right]$ and $\left[\tau_{off}^{lowlimit}, \tau_{off}^{highlimit}\right]$, respectively. Then, for each pair of $[\tau_{on}(i), \tau_{off}(i)]$, a set of RCQ parameters is calculated. Usually, when the measurement data is available, the PDFs of the ON/OFF times of the HVACs are treated as known, so the second approach is often used for modeling HVAC control groups with abundant historical operation data.

We will use two examples to illustrate the two approaches. Interested readers are encouraged to read [5] for more details regarding the randomization process.

Example 2.4 Assume that the SS-RCQ model parameters are uniformly distributed $\pm$2.5% around the values given in Table 2.2. Calculate the ON/OFF time τ_{on} and τ_{off} of a group of 1000 HVAC units when the outdoor temperature $T_{outdoor}$ is 0 °C. Plot the distribution of τ_{on} and τ_{off}.

Solution:

The MATLAB code, example 2_4.m, is used for randomizing the SS-RCQ models.

As shown in Fig. 2.8, the distribution of the ON/OFF times of the 1000 HVAC units are very close to the normal distribution, making them suitable generic HVAC model parameters for large-scale simulations.

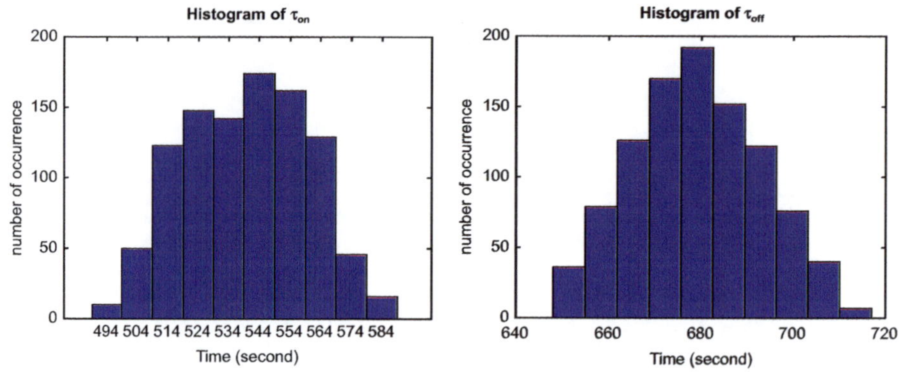

Fig. 2.8 Histogram of the ON/OFF times of 1000 HVAC units when randomly selecting the SS-RCQ model parameters from the value given in Table 2.2 by ±2.5% ($T_{outdoor} = 0$ °C)

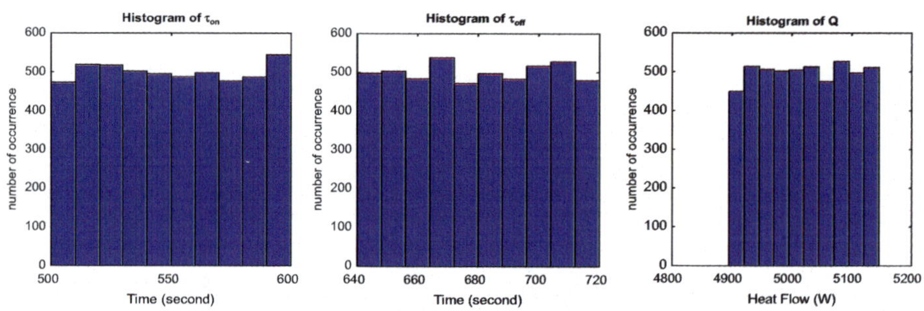

Fig. 2.9 Histograms of τ_{on} and τ_{off} and Q ($T_{outdoor} = 0$ °C)

Example 2.5 Assume that the ON/OFF time τ_{on} and τ_{off} of a group of 1000 HVAC units are uniformly distributed within 500–600 s and 640–720 s, respectively. Assuming the Q is uniformly distributed between 4895 and 5145 W, calculate the RCQ parameters using (2.4) and (2.5).

Solution:

The MATLAB code, example 2_5.m, is used for randomizing the SS-RCQ models.

Remarks

As shown in Figs. 2.9 and 2.10, the distribution of the RC parameters of the 1000 HVAC units are sort of following the normal distribution. If an actual distribution is used, then this method can effectively generate a group of RCQ parameters for the HVAC units to forecast their behaviors.

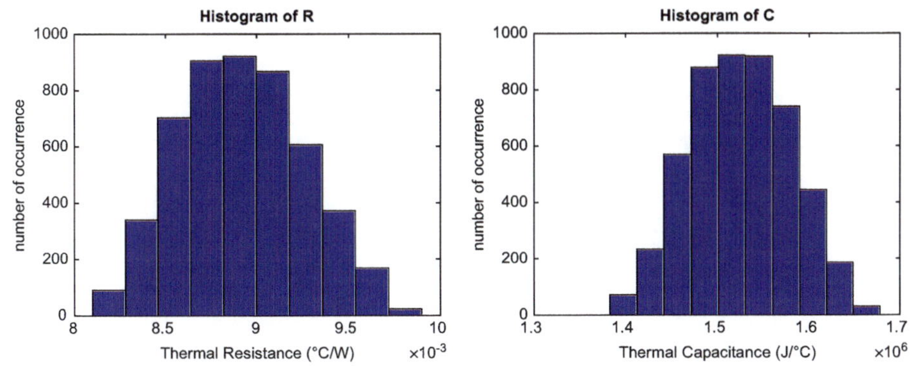

Fig. 2.10 Histogram of the RC parameters of 1000 HVAC units when τ_{on} and τ_{off} are uniformly distributed within 500–600 s and 640–720 s, respectively ($T_{outdoor} = 0$ °C)

2.6 Modeling Non-thermostatically Controllable Loads

Non-TCLs include washers, dryers, dishwasher, pool pumps, and some cooking loads such as rice cookers and slow cookers. Those loads can be put on standby so they can be turned on when needed. Note that interrupting the operation of appliances such as washers for temporary load reduction is not preferred. This is because manufacturers have optimized the operation of the washer cycles, so intermittent operation may cause suboptimal operation of the washer or shorten its lifetime.

The first modeling consideration is that electricity consumptions of most appliances are not constant over the entire operation period. As shown in Fig. 2.11, when a washer is in operation, it consumes more energy in the "spinning and stir" cycle. Usually, when the aggregated consumption of TCLs is of concern, using the average power consumption of a non-TCL load is sufficient. However, when modeling a home energy management system, because the power limit constraint needs to be met at each operation interval, the power variations at different operation stages and the transient response of an appliance will need to be accounted for.

The second modeling consideration is the start time and operation duration. Normally, the operation statistics can be derived from the historical data. For example, the PDFs of the appliance operation time and duration can be obtained, so the Monte Carlo simulation approach can be used to simulate the appliance operation for typical days in a year.

In Example 2.6, we will use the modeling of the washer and dryer loads as an example to illustrate how the non-TCL, controllable load can be modeled.

Example 2.6 Assume that the rated power of the washer is evenly distributed between 300 and 500 W and the rated power of the dryer is evenly distributed between 3000 and 4000 W. The PDF of the washing activity during a week is given in Table 2.4. The PDFs of the washer operation time over a period of 24 h on weekdays or weekends are listed in Table 2.5 and plotted in Fig. 2.12. The PDF for

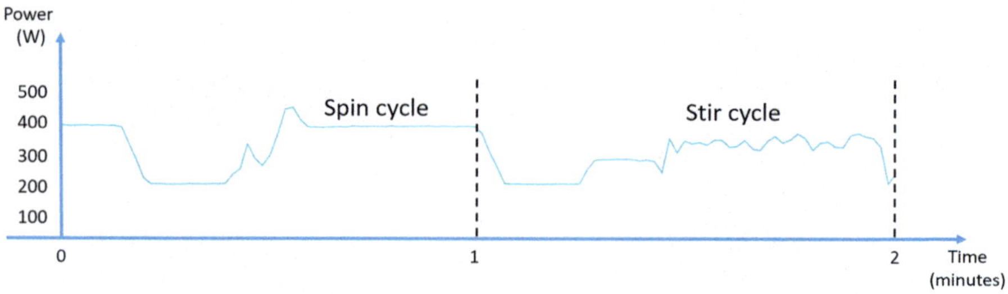

Fig. 2.11 Electricity consumption of a washer in its spin and stir cycles

Table 2.4 Probability of the washing activity for weekday and weekends

Date	Monday	Tuesday	Wednesday	Thursday	Friday	Saturday	Sunday
Probability	0.1	0.1	0.1	0.05	0.05	0.3	0.3

Table 2.5 Probability of the washer to be turned on at each hour of the day

Hour of the Day	**1**	**2**	**3**	**4**	**5**	**6**	**7**	**8**
Weekend (probability)	0	0	0	0	0	0	0	0.1
Weekday (probability)	0	0	0	0	0	0	0	0.01
Hour of the day	**9**	**10**	**11**	**12**	**13**	**14**	**15**	**16**
Weekend (probability)	0.1	0.1	0.1	0.07	0.05	0.04	0.03	0.02
Weekday (probability)	0.01	0.1	0.1	0.06	0.06	0.06	0.06	0.06
Hour of the day	**17**	**18**	**19**	**20**	**21**	**22**	**23**	**24**
Weekend (probability)	0.03	0.04	0.06	0.15	0.08	0.02	0.01	0
Weekday (probability)	0.06	0	0.06	0.1	0.1	0.1	0.06	0

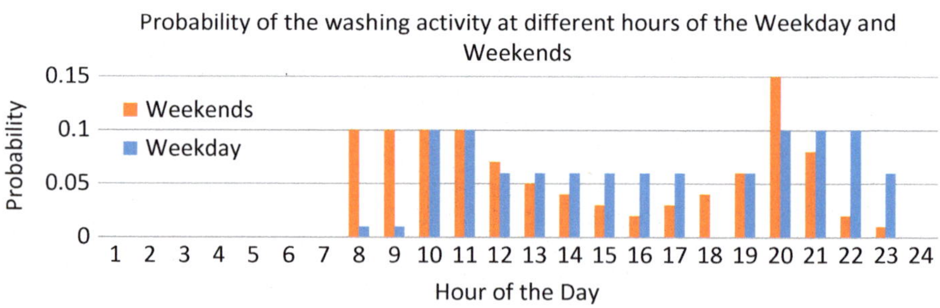

Fig. 2.12 Electricity consumption of a washer in its spin and stir cycles

the number of loadings are shown in Table 2.6. The ON/OFF time τ_{on} and τ_{off} of a group of 1000 washer and dryer pairs in 1000 households are uniformly distributed within 30–60 min and 45–75 min, respectively. The dryer will be turned on within 5–30 min after one loading is finished following the PDFs shown in Table 2.7 after a

Table 2.6 Probability of the number of loadings

Time (minute)	1	2	3	4	5	6
Probability	0.1	0.2	0.35	0.3	0.04	0.01

Table 2.7 Probability of the dryer to be turned on after one load of laundry is finished

Time (minute)	5	10	15	20	25	30
Probability	0.5	0.25	0.15	0.07	0.02	0.01

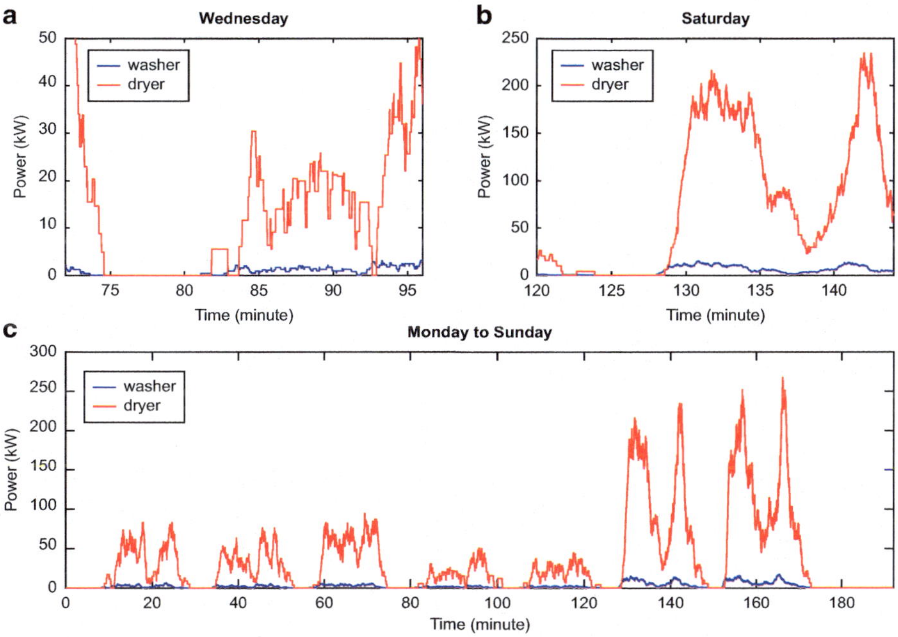

Fig. 2.13 Electricity consumptions of 1000 washer and dryer units on (**a**) Wednesday, (**b**) Saturday, and (**c**) Monday to Sunday

load of laundry is finished. If there are multiple loadings, the washer will be turned on for the next loading randomly within 5 min before or after the dryer is turned on. Plot the aggregated electricity consumption profile of the 1000 washer and dryer units for a week on 1-min resolution.

Solution:
The MATLAB code, example 2_6.m, is used for generating the following load profiles for the 1000 washer and dryer units (Fig. 2.13).

Remarks:

Because the load profiles were generated using random numbers based on PDFs, the resultant load profiles are not unique. The washer and dryer are used more frequently during weekends, so in Table 2.4, weekends have higher probabilities to be selected for washing activities. When to wash clothes and how many loadings will be determined by the probability distribution listed in Tables 2.5 and 2.6. Those PDFs can either be obtained using smart meter data or be extracted from customer survey results.

To refine this model, one can also assign probability to complete all loading in 1 day or multiple days. Furthermore, one can use measured dryer and washer load profiles (see Fig. 2.11) in the simulation instead of using the average power over the entire operation period as did in this example.

2.7 Modeling the Base Load

Base loads include standby, entertainment, lighting, cooking, and some plug-in loads. Those loads usually stay on for hours and the users normally are not willing to put off the consumptions.

Many loads, such as dishwashers, TVs, stereos, computers, stoves, and refrigerators, have standby loads of a few watts. Thus, in a house, even when all the appliances are turned off, there will be around 40–100 W consumed by those standby loads.

Entertainment loads, such as TV, stereo, and play stations, consume approximately a few hundred watts. The consumption of entertainment loads are usually well correlated with occupant behaviors and are relatively predictable. This is because once those devices are turned on, they will be on for hours.

Interior lighting loads range from tens of watts (e.g., LED lights) to a few hundred watts (e.g., incandescent lights). Exterior lighting loads are usually a few hundred watts. The lighting loads have seasonal patterns and are well correlated with daylights pattern.

Cooking loads include range, oven, microwave, toasters, slow cookers, coffee makers, and other kitchen loads. In general, cooking loads are in the range of hundreds up to 1.2 kW and they are much higher than the other types of base loads. The cooking load occurs normally within a fixed period of time for weekday and weekends. They normally last from minutes (microwave) to hours (slow cooker). Some of the cooking loads have a timer and can be controlled. Most of the time, customers may allow a cooking task to be completed early but not later than their expected finish time.

Modeling base load is extremely important when modeling demand response programs. For example, if the household is under the time-of-use rate with demand

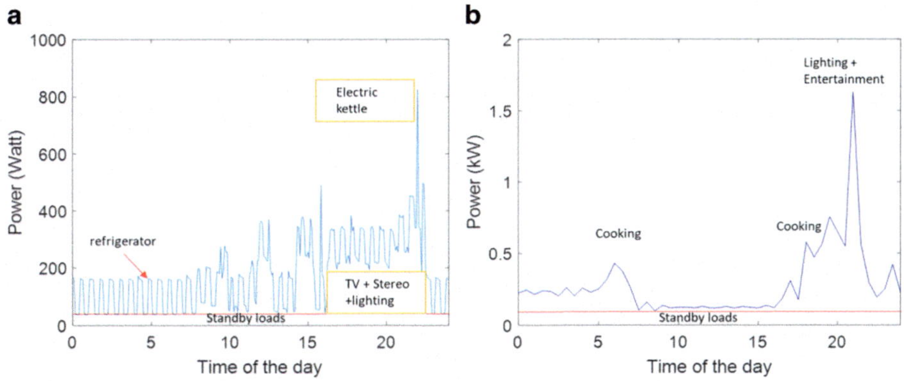

Fig. 2.14 Base loads at residential households. (**a**) Sampling rate at 5 min without cooking load and thermostatically controlled loads. (**b**) Sampling rate at 30 min with cooking load but without thermostatically controlled loads

charges,[1] the controller needs to forecast the peak of the base load in order to avoid turning on the controllable loads during those periods. The base loads are highly volatile because different person has different consumption patterns. Even for the same person, the consumption pattern may vary day to day.

As shown in Fig. 2.14a, if high resolution smart meter data (e.g., sampling rates less than 5 min) are available, one can easily identify the on/off cycles of major appliances and determine the model parameters for the baseload for typical day types.

As shown in Fig. 2.14b, if only low-resolution smart meter data (e.g., sampling rates more than 30 min) are available, sophisticated load disaggregation methods are needed to extract energy consumption patterns of the base load. A straightforward method is to subtract from the total loads the controllable loads and thermostatically controlled loads to extract the base load. This method will result in errors because the turn-on time and the turn-off time of the modeled appliances may not be well aligned with those of the actual appliances.

Example 2.7 In Fig. 2.15, what is the amount of the stand-by load? What are the information you can derive from the load profiles in weekdays? What are the information you can derive from the load profiles in weekends? What are the baseload in this house?

[1]Demand charges are designed to reduce the customer consumption during peak load periods. For example, the highest 15-minute power consumption of a month will be recorded and be charged at a much higher $/kW rate than then energy charges.

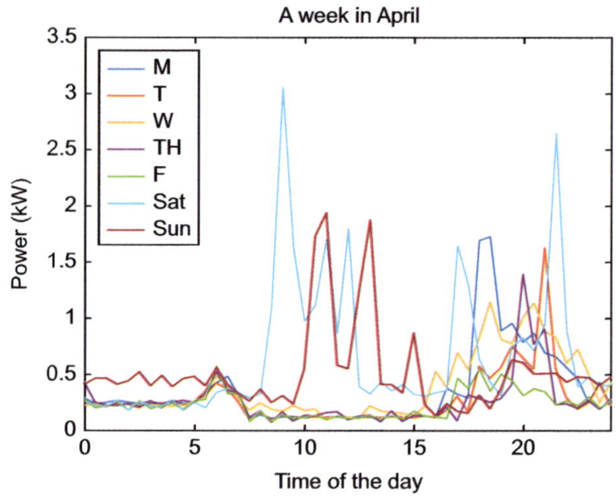

Fig. 2.15 Load profiles for a single-family house. Data length: one week in April 2016. Sampling rate: 30 min

References

1. Lu, N. (Sept. 2012). An evaluation of the HVAC load potential for providing load balancing service. *IEEE Transactions on Smart Grid, 3*(3), 1263–1270.
2. Du, P., & Lu, N. (2011). Appliance commitment for household load scheduling. *IEEE Transactions on Smart Grid, 2*(2), 411.
3. Kondoh, J., Lu, N., & Hammerstrom, D. J. (2011). An evaluation of the water heater load potential for providing regulation service. *IEEE Transactions on Power Apparatus and Systems, 26*(3), 1309–1316.
4. Vanouni, M., & Lu, N. (2015). Improving the centralized control of thermostatically controlled appliances by obtaining the right information. *IEEE Transactions on Smart Grid, 6*(2), 946.
5. Vanouni, M., & Lu, N. (2018). A reward allocation mechanism for thermostatically controlled loads participating in intra-hour ancillary services. *IEEE Transactions on Smart Grid, 9*(5), 4209–4219.

Chapter 3
Basic Control Approach for Aggregated Demand Response Programs

3.1 Direct Load Control Algorithms

DLC refers to centralized control methods that utilities or load aggregators use to directly turn on and off a controllable load remotely through wired or wireless communication networks, regardless of the customers' willingness for participating or not at the time. Thus, DR programs using DLC approach usually require the utility or load aggregator to reach an agreement with the customer who owns the load beforehand.

3.1.1 General Design Considerations

Because DLC requires customers to yield the control of their devices to a central controller, load aggregators or utilities need to provide strong financial incentives to attract customers to participate the DLC-based DR program. Therefore, DLC-based DR programs are often used to provide high-value grid services.

Load balancing services (e.g., regulation and load following) are ancillary services that require control signals being followed accurately on a timely manner. In an electricity market, the size of the load balancing market is usually within a few hundred MWs and load balancing service prices are normally higher than energy service prices. Therefore, in the following subsections, we will use the design of DLC control algorithms for providing load balancing service as an example to illustrate the general design principles.

Suitable load-side resources for DLC-based DR programs need to have the following characteristics. *First*, the DR resource needs to be in operation frequently and the duration of the "on" and "off" periods need to be longer than 5 min. Because load balancing signals are issued every few seconds, resources operate once or twice a day with very short "on" and "off" durations are not worthy of consideration.

P. Du et al., *Demand Response in Smart Grids*,
https://doi.org/10.1007/978-3-030-19769-8_3

Second, the DR resource needs to have a high power capacity to obtain an appreciable response with few appliances. *Third*, the DR resource needs to have a broad comfort setting. For example, an HVAC unit with a wider thermostat dead band setting will be a better resource than the one with a narrower dead band setting.

By controlling the on/off of the load, DLC allows the central DR controller to adjust the aggregated loads precisely. However, it will be hard to gain customer acceptance if the control of their devices will compromise their comfort significantly. Thus, the design of a control algorithm that meets grid service requirements without compromising customer comfort is crucial for customer satisfaction and retention.

The design of DLC also needs to account for communication network problems. To verify the participation and quantify response, bidirectional communication links are often needed between the central controller and hundreds or thousands of DR resources over a vast geographic area. Among access technologies, wireless communication is the most common solution because of its low cost and vast coverage. Therefore, it is critical to understand the reliability of wireless communications and quantify its impact on DR performance, especially on the DLC-based DR program that requires frequent information exchange between the controller and end devices.

Lastly, because many DR program targets end use loads connected to the grid from the distribution side, the impact of DR on distribution system operation needs to be addressed if DR penetration is higher than 20%. This is because 20% load swings at distribution transformers and substations may cause equipment overload, under-voltage, or over-voltage issues under certain operation conditions.

3.1.2 Test System Setup

A test system that consists of one thousand 6 kW HVAC units in its heating mode is set up to illustrate the development of a DLC control algorithm for providing load balancing services. The simplified RCQ model introduced in Sect. 3.3.1 is used to model the physical operation of the 1000 HVAC units. Parameter randomization introduced in Sect. 3.5 is used to generate different operation characteristics for each HVAC unit. As shown in Fig. 3.1, the thermostat high limit, T^{+}, is 23 °C with a dead band ($deadband = T^{+} - T^{-}$) of 2 or 4 °C. Therefore, thermostat low limit, T^{-}, is either 21 °C or 19 °C. The mean values of C, R, and Q are set to 3599.3 J/°C, 0.1208 °C/W, and 400 W, respectively. The R, C, and Q parameters for the HVAC units are randomized for different HVAC units to create load diversity. As shown in Fig. 3.2, six outdoor temperature profiles with daily average temperatures, T_{ave}, ranging from −10 to 15 °C, are used to model different weather conditions.

The central controller is equipped with a forecaster to estimate the room temperature for the next time step, determine the on/off status of the HVAC units, and

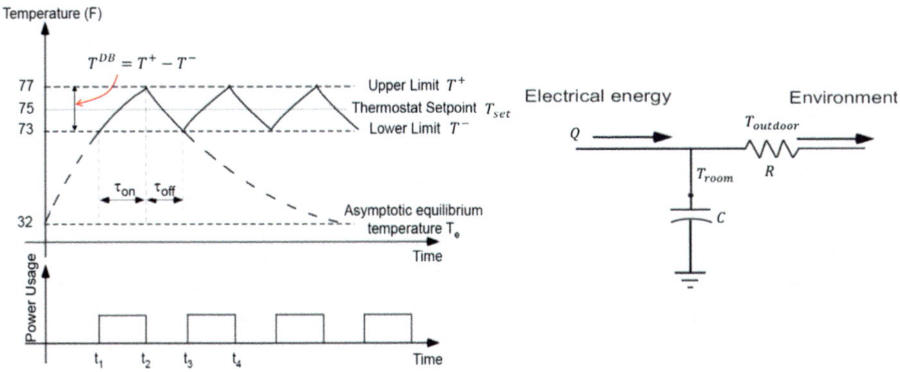

Fig. 3.1 Room temperature profile, HVAC on/off curve, and equivalent electric circuit model of a residential HVAC system

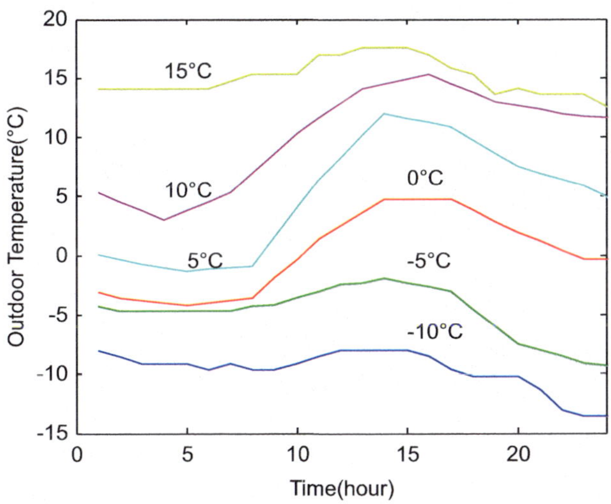

Fig. 3.2 Outdoor temperature profiles

create two priority (turn-on and turn-off) lists for the two groups of HVAC units. The forecaster uses the linearized-measurement-based ETP model introduced in Sect. 3.3.2 to forecast the HVAC operation. As shown in Fig. 3.3, the forecasted room temperature, T_f^{HVAC}, is tuned by the actual measurement, T_a^{HVAC}, which is collected from each HVAC unit every 15 min. This forecast-and-update process will reduce data traffic between HVAC units and the central controller. For more detailed analysis on the impact of the update-by-measurement process, please refer to Sect. 3.4 and references [1, 2].

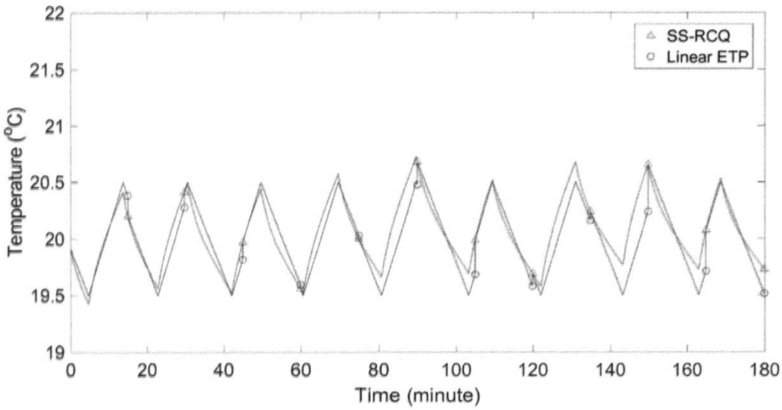

Fig. 3.3 Space heating unit behavior modeled by the simplified model and tuned by measurements

3.1.3 Test System Initialization

An initialization process is normally necessary to obtain unbiased initial conditions for the HVAC control group. This is because if the initial room temperature and on/off status of each HVAC unit are randomly selected assuming uniform distributions, there are approximately 50% HVAC units in the "on" group and the room temperature of each house will be evenly distributed within the given temperature dead band. However, in winter, there will be more than 50% units in the "on" group if the outdoor temperature is below −15 °C and less than 50% units in the "on" group if the outdoor temperature is above 15 °C.

The initialization includes three steps. First, randomly set the initial room temperature of each HVAC unit to be a value between T^- and T^+. Second, randomly set the status of each HVAC unit to be either "on" or "off." Third, take the first value of the 24-h outdoor temperature profile, run the simulation for 10 h. At the end of the 10 h, record the on/off status and room temperature of each HVAC unit and use them as the initial values of the subsequent simulation.

The initialization process of a load group consisting of five HVAC units is shown in Fig. 3.4. At the beginning of the initialization process, actions of a few HVAC units are synchronized. But after the simulation runs for 10 h with $T_{outdoor} = 5\text{C}$, we can see that both the on/off status and the room temperatures of the HVAC units have been randomized.

3.1.4 Construction of Operation Baseline

An aggregated baseline output of the HVAC units, $P_{baseline}$, must be provided to grid operators so that deviations from $P_{baseline}$ can be measured as load balancing services (e.g., regulation up and regulation down services).

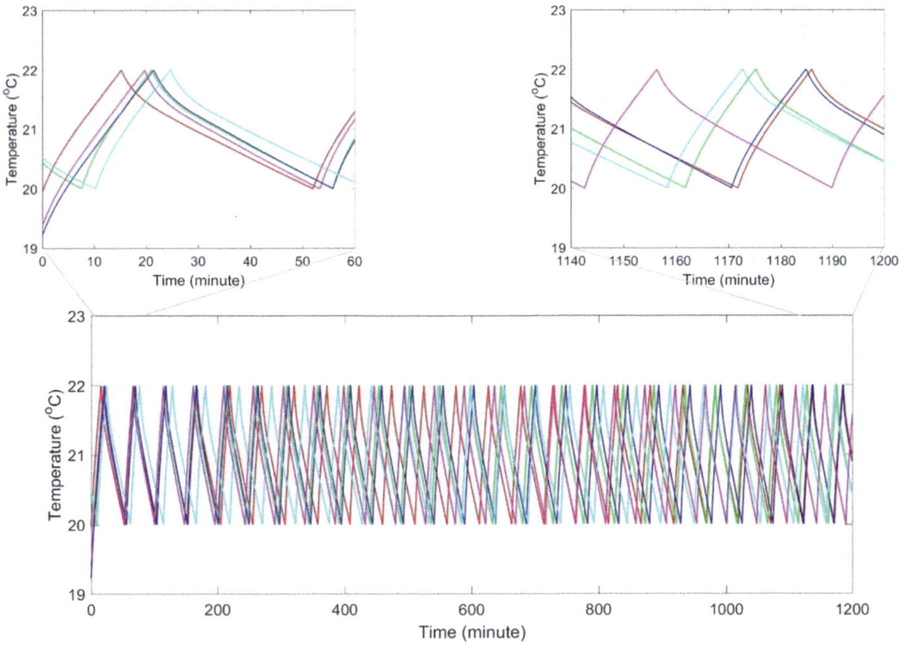

Fig. 3.4 HVAC initialization process ($T_{outdoor} = 5^{\circ}$C)

To create a baseline load, all participating HVAC units are modeled in an uncontrolled mode using next-day outdoor temperature forecasts. The uncontrolled model means that an HVAC is turned on and off based on its thermostat setting. Then, the aggregated HVAC power outputs (black lines in Fig. 3.5) are averaged to an hourly load profile as the day-ahead $P_{baseline}$ (red lines in Fig. 3.5). Note that the baseline load profiles vary with different thermostat settings and different numbers of controlled HVAC units, as illustrated in Figs. 3.6 and 3.7. In general, more HVAC units and wider dead band settings provide greater load balancing capacity.

3.1.5 Construction of the Control Signal

The 4-s area control error (ACE) signal, P_{ACE}, and load following signal, P_{LF}, are down-sampled to 1 min. Then, magnitudes of the ACE and load following signals are scaled to ±0.2, ±0.6, and ± 1 MW. The scaled signals will be added up to the forecasted HVAC baseline load obtained in Sect. 3.1.4 to get the control signals, P_c^{LF} and P_c^{ACE}, which are calculated as:

$$P_C^{LF} = P_{baseline} + P_{LF} \tag{3.1}$$

$$P_C^{ACE} = P_{baseline} + P_{ACE} \tag{3.2}$$

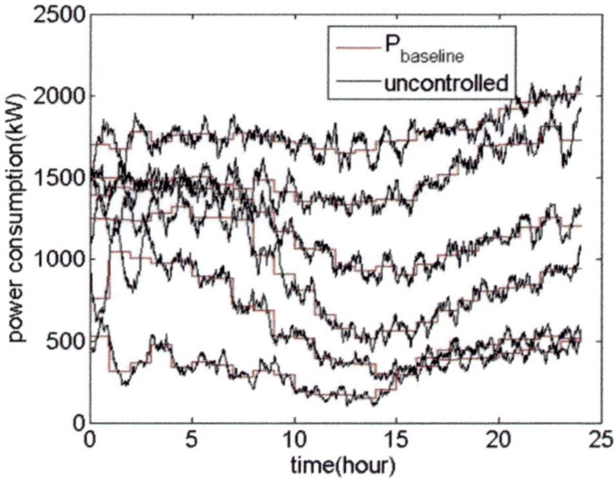

Fig. 3.5 Construction of HVAC baseline loads using six sets of outdoor temperature profiles (temperature profiles: $T_{\text{ave}} = -10, -5, 0, 5, 10, 15\ ^\circ\text{C}$)

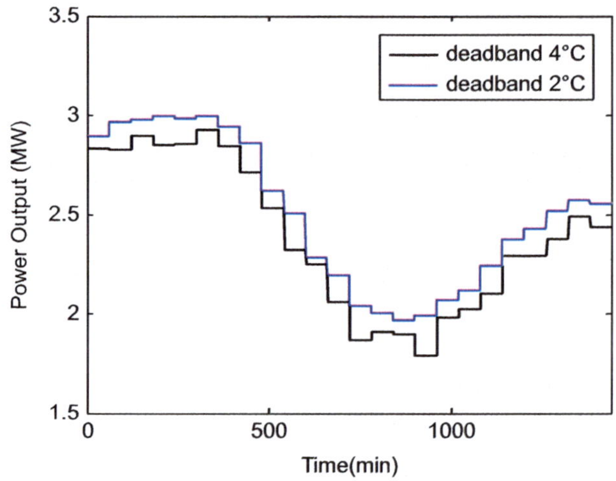

Fig. 3.6 Construction of hourly HVAC baseline loads using different temperature dead bands ($T_{\text{ave}} = 0\ ^\circ\text{C}$; 1000 HVAC units)

Examples of load following and ACE control signals are shown in Fig. 3.8. Three load following signals with different magnitudes overlaying with the HVAC-based load profile are shown in Fig. 3.9.

3.1.6 Priority-List-Based HVAC Control Algorithm

A flow chart of the direct load control logic is provided in Fig. 3.10.

The 1000 HVAC 6 kW units are divided into two groups based on their on/off status. As the space heating mode is used in the simulation, the units in the "on" group are prioritized in descending order based on their room temperatures, i.e., if

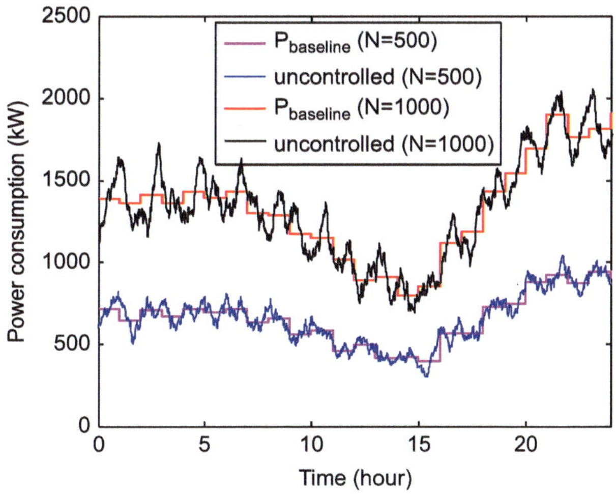

Fig. 3.7 Construction of hourly HVAC baseline loads using different number of HVAC units ($T_{\text{ave}} = 0\ °\text{C}$)

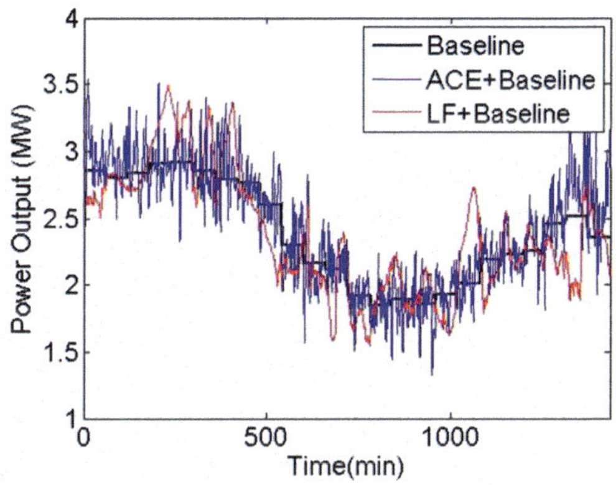

Fig. 3.8 Construction of control signals ($T_{AVE} = 0^{\circ}\text{C}$, 1000 HVAC units)

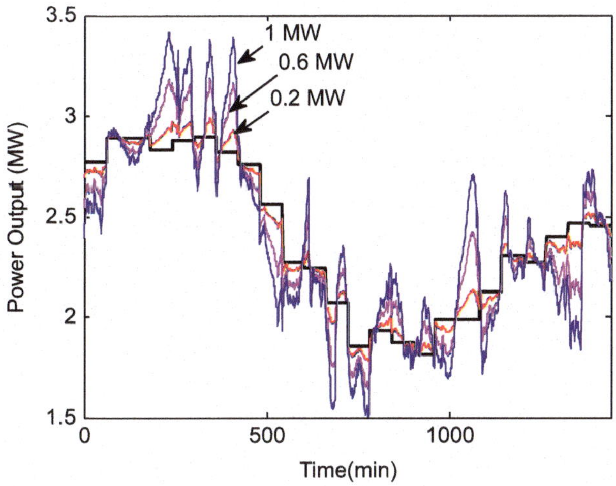

Fig. 3.9 Different load following signals

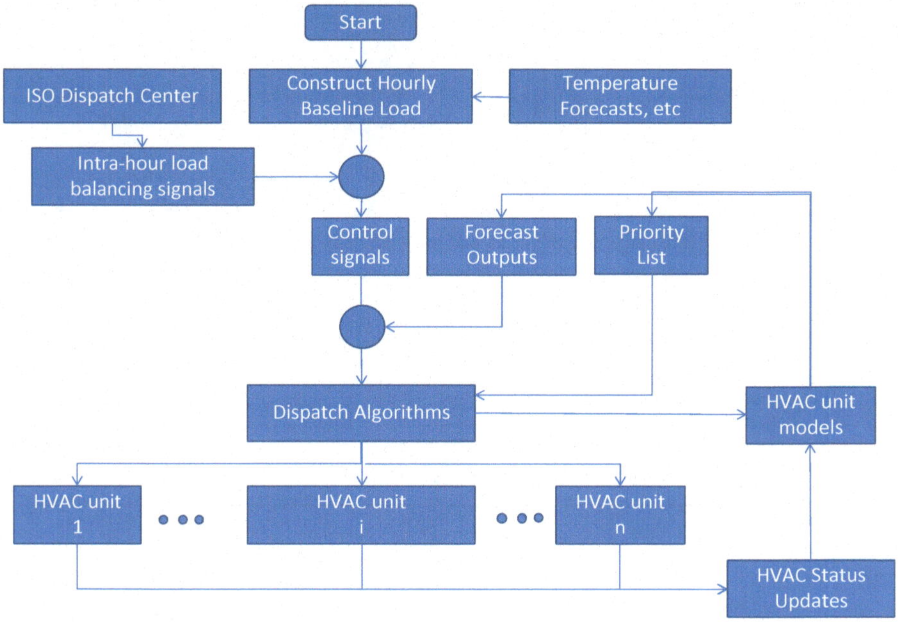

Fig. 3.10 Flow chart of HVAC control logic

the room temperature is closer to the upper thermostat setting T^{+}, the unit is at the top of the queue to be turned off. The units in the "off" group are prioritized in ascending order based on their room temperatures, i.e., if the room temperature is closer to the lower thermostat setting T^{-}, the unit is at the top of the queue to be turned on. The HVAC units that are "on" under direct load control will switch off immediately when they receive an "off" signal from the central controller, and vice versa.

Let's use a simple example to illustrate the operation of the "no-controlled," "baseline," and "load balancing" cases.

3.1.6.1 No-Controlled Case

Assume that 10 HVAC units have the same thermostat setting, i.e., setpoint at 70 °F with a dead band of 4 °F. Each HVAC unit consumes 6 kW when turned on. Operation status of the 10 HVAC units at time $t = 1$ is listed in Table 3.1. As can be seen in the table, there are 5 "on" units and 5 "off" units.

Room temperatures of each house for the no-control case during the first 30 min are listed in Table 3.2. As can be seen from the results, if each HVAC unit is controlled by the thermostat, the total power consumption of the 10 HVAC units will vary between 24 and 30 kW.

Table 3.1 Ten appliance with different room temperature and on/off status at time *t*

Appliance #	1	2	3	4	5	6	7	8	9	10
Room temperature (°F)	71.5	70.5	69.5	68	71.9	70.2	68.6	71.2	69.9	70
On/off status	On	On	Off	On	Off	On	Off	Off	On	Off

Table 3.2 Room temperature and on/off status of 10 HVAC units without any control in a 30-min interval (Temperature is in °F)

Time (minute)	1	2	3	4	5	6	7	8	9	10	Power (kW)
1	71.5	70.5	69.5	68	71.9	70.2	68.6	71.2	69.9	70	30
2	71.8	70.65	69.37	68.3	71.57	70.64	68.45	70.85	70.1	69.8	30
3	72.1	70.8	69.24	68.6	71.24	71.08	68.3	70.5	70.3	69.6	30
4	71.76	70.95	69.11	68.9	70.91	71.52	68.15	70.15	70.5	69.4	24
5	71.42	71.1	68.98	69.2	70.58	71.96	68	69.8	70.7	69.2	24
6	71.08	71.25	68.85	69.5	70.25	72.4	68.41	69.45	70.9	69	30
7	70.74	71.4	68.72	69.8	69.92	72.15	68.82	69.1	71.1	68.8	24
8	70.4	71.55	68.59	70.1	69.59	71.9	69.23	68.75	71.3	68.6	24
9	70.06	71.7	68.46	70.4	69.26	71.65	69.64	68.4	71.5	68.4	24
10	69.72	71.85	68.33	70.7	68.93	71.4	70.05	68.05	71.7	68.2	24
11	69.38	72	68.2	71	68.6	71.15	70.46	67.7	71.9	68	24
12	69.04	71.68	68.07	71.3	68.27	70.9	70.87	68.07	72.1	68.3	24
13	68.7	71.36	67.94	71.6	67.94	70.65	71.28	68.44	71.75	68.6	24
14	68.36	71.04	68.33	71.9	68.31	70.4	71.69	68.81	71.4	68.9	30
15	68.02	70.72	68.72	72.2	68.68	70.15	72.1	69.18	71.05	69.2	30
16	67.68	70.4	69.11	71.97	69.05	69.9	71.95	69.55	70.7	69.5	24
17	67.98	70.08	69.5	71.74	69.42	69.65	71.8	69.92	70.35	69.8	30
18	68.28	69.76	69.89	71.51	69.79	69.4	71.65	70.29	70	70.1	30
19	68.58	69.44	70.28	71.28	70.16	69.15	71.5	70.66	69.65	70.4	30
20	68.88	69.12	70.67	71.05	70.53	68.9	71.35	71.03	69.3	70.7	30
21	69.18	68.8	71.06	70.82	70.9	68.65	71.2	71.4	68.95	71	30
22	69.48	68.48	71.45	70.59	71.27	68.4	71.05	71.77	68.6	71.3	30
23	69.78	68.16	71.84	70.36	71.64	68.15	70.9	72.14	68.25	71.6	30
24	70.08	67.84	72.23	70.13	72.01	67.9	70.75	71.79	67.9	71.9	30
25	70.38	67.99	72.1	69.9	71.68	68.34	70.6	71.44	68.1	72.2	30
26	70.68	68.14	71.97	69.67	71.35	68.78	70.45	71.09	68.3	72	24
27	70.98	68.29	71.84	69.44	71.02	69.22	70.3	70.74	68.5	71.8	24
28	71.28	68.44	71.71	69.21	70.69	69.66	70.15	70.39	68.7	71.6	24
29	71.58	68.59	71.58	68.98	70.36	70.1	70	70.04	68.9	71.4	24
30	71.88	68.74	71.45	68.75	70.03	70.54	69.85	69.69	69.1	71.2	24

Boxes highlighted in green represent the "on" periods and in yellow "off" periods

3.1.6.2 Baseline Case

Assume that the baseline consumption of this 10-HVAC load control group is 30 kW such that there are always 5 "on" units and 5 "off" units. Use the same initial condition as the no-controlled case. Turn-on and turn-off priority lists at time $t = 1$ are shown in Fig. 3.11.

	On list	Off list
71-72°F	1	
70-71°F	2,6	5,8,10
69-70°F	9	3
68-69°F	4	7

highest priority | Medium priority | Lowest priority | Locked

Fig. 3.11 Turn-on Priority list and Turn-off priority list of the 10 HVAC unit listed in Table 3.1 at time *t*

Room temperatures of each house for the "baseline" case are listed in Table 3.3. If the central controller is controlling the 10 HVAC units to follow a baseline of 30 kW, at minutes 4, 7, 16, 21, 24, 25, and 26, some HVAC units will be turned on earlier and at minutes 12 and 14, some HVAC units are turned off earlier. As a result, the total power consumption of the 10 HVAC units will be maintained at 30 kW.

In Table 3.3, we can also observe the advantage of using the room temperature-based priority list to rank the units. When letting units that are closest to the end of its on/off cycle to be turned off/on, we can reduce the number of switching for each HVAC unit and maximize the duration for each unit to stay in the same operation mode. This will allow the load following service to be provided without shortening HVAC lifetimes and violating customer room temperature settings. This example, though simple, is representative for controlling a group of HVAC units to maintain a constant baseline over a period of time.

3.1.6.3 Load Following Case

As shown in Table 3.4, after receiving a load balancing signal, the controller will again follow the priority list to select units to be turned on and off so that the total power consumption will equal to the baseline plus the load balancing signal. In the "power" column, cells highlighted in "red" are time instances providing regulation-up signals and in "blue" are time instances providing regulation-down signals. HVAC units will be turned off/on before the room temperatures reach its upper/lower limits, respectively, to consume less or more power compared with the baseline energy consumption at 30 kW.

There are a few observations:

- The minimum power resolution is determined by the HVAC rated power because we cannot turn an HVAC unit half-on or half-off. When providing load balancing service at MW level, this would not be an issue. However, when providing load balancing service to a microgrid with several hundred kW loads, one may need to evaluate whether or not such accuracy is sufficient.
- The room temperature each HVAC unit regulates will maintain within the bandwidth if the load balancing signal is energy neutral. Otherwise, when room temperatures of all houses drop/rise close to the lower/upper temperature limits,

Table 3.3 Room temperature and on/off status of 10 HVAC units with DLC control following a 30 kW baseline operation in a 30-min interval (Temperature is in °F)

Time (minute)	1	2	3	4	5	6	7	8	9	10	Power (kW)
1	71.5	70.5	69.5	68	71.9	70.2	68.6	71.2	69.9	70	30
2	71.8	70.65	69.37	68.3	71.57	70.64	68.45	70.85	70.1	69.8	30
3	72.1	70.8	69.24	68.6	71.24	71.08	68.3	70.5	70.3	69.6	30
4	71.76	70.95	69.11	68.9	70.91	71.52	68.71	70.15	70.5	69.4	30
5	71.42	71.1	68.98	69.2	70.58	71.96	69.12	69.8	70.7	69.2	30
6	71.08	71.25	68.85	69.5	70.25	72.4	69.53	69.45	70.9	69	30
7	70.74	71.4	69.24	69.8	69.92	72.15	69.94	69.1	71.1	68.8	30
8	70.4	71.55	69.63	70.1	69.59	71.9	70.35	68.75	71.3	68.6	30
9	70.06	71.7	70.02	70.4	69.26	71.65	70.76	68.4	71.5	68.4	30
10	69.72	71.85	70.41	70.7	68.93	71.4	71.17	68.05	71.7	68.2	30
11	69.38	72	70.8	71	68.6	71.15	71.58	67.7	71.9	68	30
12	69.04	71.68	71.19	71.3	68.27	70.9	71.99	68.07	71.55	68.3	30
13	68.7	71.36	71.58	71.6	67.94	70.65	72.4	68.44	71.2	68.6	30
14	68.36	71.04	71.97	71.9	68.31	70.4	72.25	68.81	70.85	68.9	30
15	68.02	70.72	72.36	72.2	68.68	70.15	72.1	69.18	70.5	69.2	30
16	68.32	70.4	72.23	71.97	69.05	70.59	71.95	69.55	70.15	69.5	30
17	68.62	70.08	72.1	71.74	69.42	71.03	71.8	69.92	69.8	69.8	30
18	68.92	69.76	71.97	71.51	69.79	71.47	71.65	70.29	69.45	70.1	30
19	69.22	69.44	71.84	71.28	70.16	71.91	71.5	70.66	69.1	70.4	30
20	69.52	69.12	71.71	71.05	70.53	72.35	71.35	71.03	68.75	70.7	30
21	69.82	68.8	71.58	70.82	70.9	72.1	71.2	71.4	68.95	71	30
22	70.12	68.48	71.45	70.59	71.27	71.85	71.05	71.77	69.15	71.3	30
23	70.42	68.16	71.32	70.36	71.64	71.6	70.9	72.14	69.35	71.6	30
24	70.72	68.31	71.19	70.13	72.01	71.35	70.75	71.79	69.55	71.9	30
25	71.02	68.46	71.06	70.43	71.68	71.1	70.6	71.44	69.75	72.2	30
26	71.32	68.61	70.93	70.73	71.35	70.85	71.01	71.09	69.95	72	30
27	71.62	68.76	70.8	71.03	71.02	70.6	71.42	70.74	70.15	71.8	30
28	71.92	68.91	70.67	71.33	70.69	70.35	71.83	70.39	70.35	71.6	30
29	72.22	69.06	70.54	71.63	70.36	70.1	72.24	70.04	70.55	71.4	30
30	72.52	69.21	70.41	71.93	70.03	69.85	72.65	69.69	70.75	71.2	30

Green boxes: "on" periods; yellow boxes: "off" periods; red boxes: "forced on"; black boxes: "forced off"

one can no longer use the HVAC units to regulate the total energy consumption without violating the room temperature constraints.

- The number of switching will increase when letting the HVAC units following a baseline consumption or responding to load balancing signals. This is because the natural heating or cooling cycles are shortened when responding to a control signal. However, as long as there are a large number of HVAC units in the load group, the adjustment of the cycling length will be minimum for each HVAC unit.

For more detailed analysis, readers are recommended to read the papers attached at the end of this chapter.

Table 3.4 Room temperature and on/off status of 10 HVAC units with DLC control following a 30 kW baseline plus a ± 6 kW load balancing signal in a 30-min interval (Temperature is in °F)

Time (minute)	1	2	3	4	5	6	7	8	9	10	Power (kW)	Reglation signal (kW)
1	71.5	70.5	69.5	68	71.9	70.2	68.6	71.2	69.9	70	30	0
2	71.8	70.65	69.37	68.3	71.57	70.64	68.45	70.85	70.1	69.8	30	0
3	72.1	70.8	69.24	68.6	71.24	71.08	68.86	70.5	70.3	69.6	36	6
4	71.76	70.95	69.63	68.9	70.91	71.52	69.27	70.15	70.5	69.4	36	6
5	71.42	71.1	70.02	69.2	70.58	71.27	69.68	69.8	70.7	69.2	30	0
6	71.08	70.78	70.41	69.5	70.25	71.02	70.09	69.45	70.9	69	24	-6
7	70.74	70.46	70.8	69.8	69.92	70.77	70.5	69.1	71.1	68.8	24	-6
8	70.4	70.14	71.19	70.1	69.59	70.52	70.91	68.75	71.3	68.6	24	-6
9	70.06	69.82	71.58	70.4	69.26	70.27	71.32	68.4	71.5	68.4	24	-6
10	69.72	69.5	71.97	70.7	68.93	70.02	71.73	68.05	71.7	68.2	24	-6
11	69.38	69.18	72.36	71	68.6	69.77	72.14	67.7	71.9	68	24	-6
12	69.04	68.86	72.23	71.3	68.27	69.52	72.55	68.07	72.1	68.3	30	0
13	69.34	68.54	72.1	71.6	68.64	69.27	72.4	68.44	71.75	68.6	30	0
14	69.64	68.22	71.97	71.9	69.01	69.02	72.25	68.81	71.4	68.9	30	0
15	69.94	67.9	71.84	72.2	69.38	69.46	72.1	69.18	71.05	69.2	36	6
16	70.24	68.05	71.71	71.97	69.75	69.9	71.95	69.55	70.7	69.5	36	6
17	70.54	68.2	71.58	71.74	70.12	70.34	71.8	69.92	70.35	69.8	36	6
18	70.84	68.35	71.45	71.51	70.49	70.78	71.65	70.29	70	70.1	36	6
19	71.14	68.5	71.32	71.28	70.86	71.22	71.5	70.66	69.65	70.4	36	6
20	70.8	68.65	71.19	71.05	71.23	71.66	71.35	71.03	69.3	70.7	30	0
21	70.46	68.8	71.06	70.82	71.6	72.1	71.2	71.4	68.95	71	30	0
22	70.12	68.95	70.93	70.59	71.97	71.85	71.05	71.77	68.6	71.3	24	-6
23	69.78	69.1	70.8	70.36	72.34	71.6	70.9	72.14	68.25	71.6	24	-6
24	69.44	69.25	70.67	70.13	72.71	71.35	70.75	71.79	68.45	71.9	24	-6
25	69.74	69.4	70.54	69.9	72.38	71.1	70.6	71.44	68.65	72.2	24	-6
26	70.04	69.55	70.41	70.2	72.05	70.85	70.45	71.09	68.85	72	24	-6
27	70.34	69.7	70.28	70.5	71.72	70.6	70.3	70.74	69.05	71.8	24	-6
28	70.64	69.85	70.67	70.8	71.39	70.35	70.15	70.39	69.25	71.6	30	0
29	70.94	70	71.06	71.1	71.06	70.1	70	70.04	69.45	71.4	30	0
30	71.24	70.15	71.45	71.4	70.73	69.85	69.85	69.69	69.65	71.2	30	0

3.1.7 *Performance for Providing Load Following Services*

The aggregated HVAC load normally follows the control signals very well for a dead band of 4 °C. The number of cycles of a heater unit ($T_{ave} = 0$ °C; dead band is 4 °C) is around 14–20 (see blue line in Fig. 3.12); to provide the 0.2–1 MW load following service, there will be 2 more cycles on average for each HVAC unit with dead bands set at 4 °C.

However, when the dead band is narrower, the capability of the HVAC load to follow the load following signal is reduced, because tracking the control signals will partially synchronize diversified HVAC loads, as shown in Figs. 3.12, 3.13, and 3.14. When T_{room} of all households is close to T^{+} or T^{-}, the continuing "on" or "off" status of the HVAC units will cause T_{room} to exceed T^{+}or T^{-}, as shown at 420 min in Subplot 3, Fig. 3.14 ($T_{room} > T^{+}$). The occurrence of the violation can be minimized by increasing the number of HVAC units, increasing T^{+}, or decreasing T^{-}. In

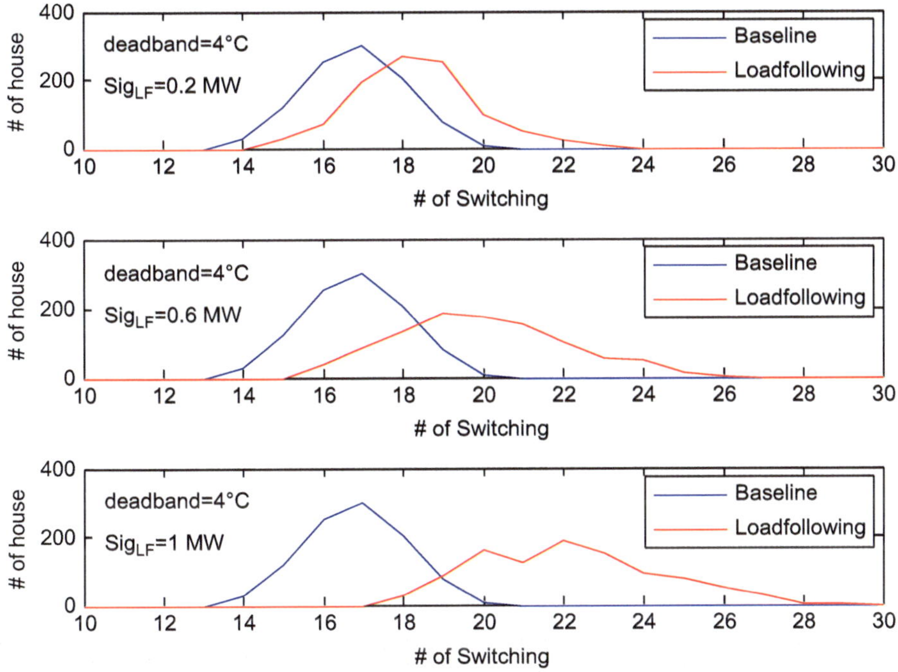

Fig. 3.12 Impact of different control signal magnitudes on HVAC daily cycles

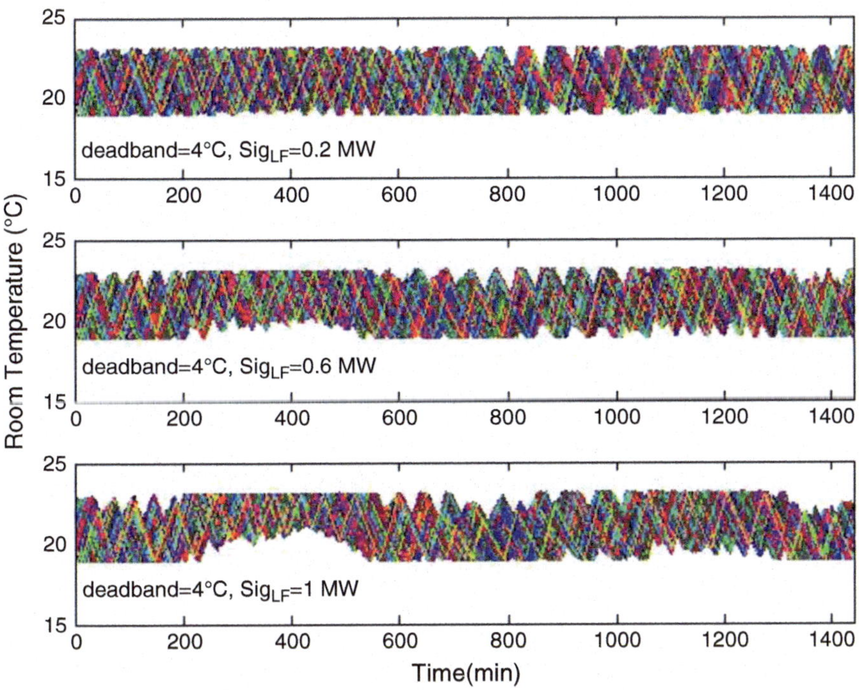

Fig. 3.13 Impact of different control signal magnitudes on room temperature profiles

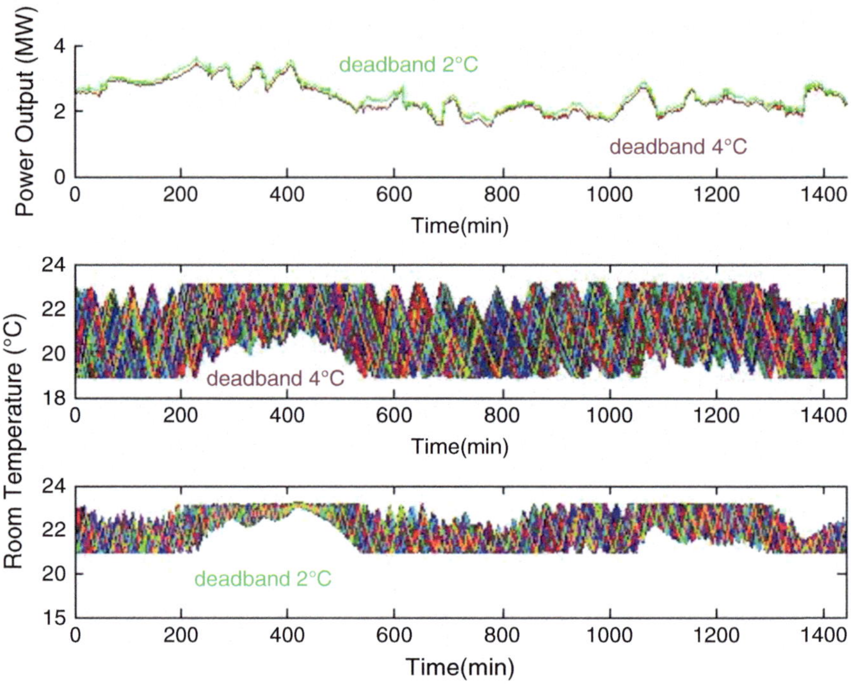

Fig. 3.14 Impact of different dead bands on room temperature profiles

addition, when the dead band is 2 °C, the HVAC unit cycles two to four times more often to follow the control signals (see Fig. 3.15), shortening the life of the unit.

When outdoor temperatures are high, the HVAC unit may be unable to provide enough load following capacity because its base load may be lower than the required load following down capacity (see the 15 °C case in Fig. 3.16). Note that the indoor temperatures in all cases (see Fig. 3.17) are always kept within their high and low limits; the centralized-dispatch algorithm simply changes the cycle length of each HVAC unit to obtain an aggregated load profile that matches the control signal.

A number of factors can limit the capacity or reduce the quality of the balancing capabilities provided by the HVAC load. For example, to avoid stalling, when an HVAC unit is turned off, a minimum turn-off time, t_{off}, needs to be considered in the HVAC unit model so that during the minimum turn-off period, the HVAC unit is locked out for the "turn-on" service, reducing the balancing capacity. In addition, the impact of random communication delays and turn-on and turn-off delays may influence how well the HVAC load can follow the control signal. The readers are advised to reference to [1–6] for how to account for those additional considerations.

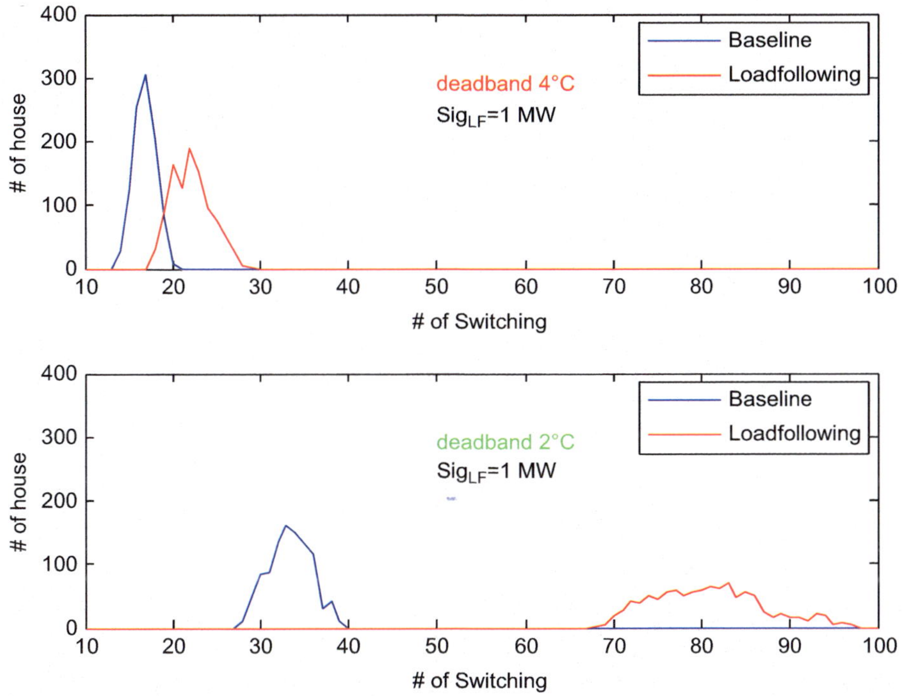

Fig. 3.15 Impact of dead bands on HVAC daily cycles

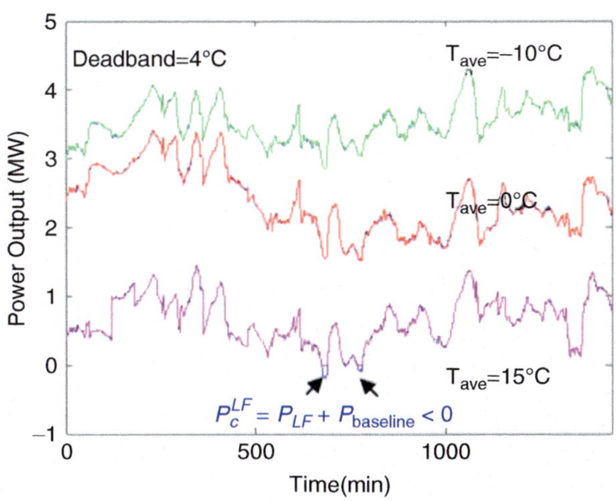

Fig. 3.16 Impact of outdoor temperatures on control signal following capabilities

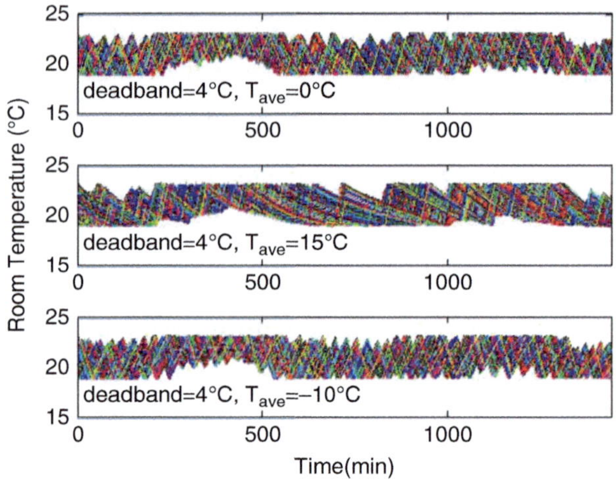

Fig. 3.17 Impact of outdoor temperatures on room temperature profiles

3.2 Indirect Load Control Algorithms

In ILC, the power consumption of the load is controlled by consumers either manually or through automated end devices (e.g., smart thermostats, timers, and smart switches). Control signals sent by a central controller are usually compounded with economic incentives, such as prices, coupons, or rebates. Because each consumer or device has its own response curve unknown to the central controller, the aggregated DR response of the ILC usually is not as accurate as that of the DLC. However, the setup of an ILC control group is much simpler and does not require two-way communication links, so it is cheaper to implement and can be widely used in reducing peak demand or shifting energy consumptions from peak hours to valley hours.

In this subsection, we will use three examples to discuss the main design considerations of the ILC approach. Three price signals will be used: time-of-use (TOU), critical peak price (CPP), and real-time price (RTP). Again, 1000 HVAC units in their heating modes will be used to illustrate the ILC-based DR programs.

3.2.1 Response Curves

ILC does not directly control the on/off of an end use load, instead, it provides an economic incentive to encourage customer participation. Thus, a response curve can be used to represent the relationship between the customer reaction and the economic incentive.

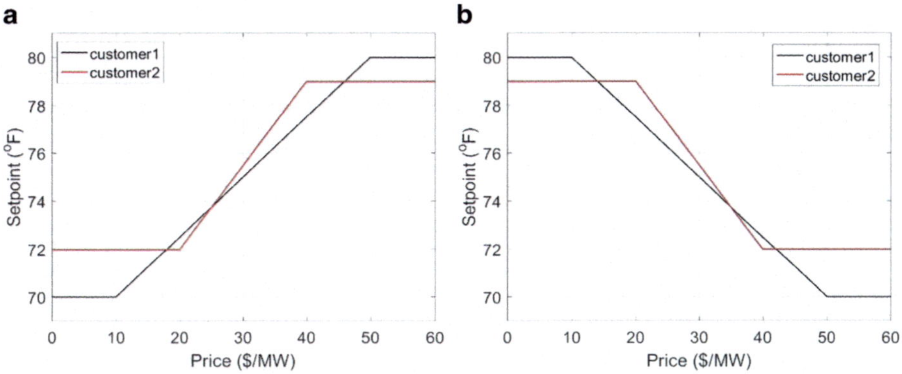

Fig. 3.18 Setpoint versus Price curves. (**a**) HVAC in cooling mode; (**b**) HVAC in heating mode

Let B be the price signal and T_{set} be the setpoint of the HVAC thermostat. By letting T_{set} respond to price signals, the following four variables will define the response curve: minimum/maximum thermostat setpoint, $\left[T^{min}_{setpoint}, T^{max}_{setpoint}\right]$, and their corresponding price signals, $[B^{min}, B^{max}]$.

As shown in Fig. 3.18a,b, the response curve of an HVAC unit in its cooling/heating modes can be represented by (3.3) and (3.4), respectively.

$$T_{set} = \begin{cases} T^{min}_{setpoint} & B < B^{\min} \\ T^{min}_{setpoint} + \dfrac{T^{max}_{setpoint} - T^{min}_{setpoint}}{B^{max} - B^{min}}\left(B - B^{min}\right) & B^{\min} \leq B \leq B^{\min} \\ T^{max} & B > B^{\max} \end{cases} \quad (3.3)$$

$$T_{set} = \begin{cases} T^{max}_{setpoint} & B < B^{min} \\ T^{min}_{setpoint} + \dfrac{T^{min}_{setpoint} - T^{max}_{setpoint}}{B^{max} - B^{min}}\left(B - B^{max}\right) & B^{min} \leq B \leq B^{min} \\ T^{min}_{setpoint} & B > B^{max} \end{cases} \quad (3.4)$$

As shown in Fig. 3.18, in summer, the first customer will respond following the black response curve defined by $[10, 50, 70, 80]$ and the second customer will respond following the red response curve defined by $[20, 40, 72, 79]$. In winter, the first customer will respond following the black response curve defined by $[10, 50, 80, 70]$ and the second customer will respond following the red response curve defined by $[20, 40, 79, 72]$, as shown in Fig. 3.18b. Note that the second customer has a narrower setpoint range (i.e., 7 °F) and price response range (i.e., \$20/MW), compared with the first customer.

A central controller can broadcast price signals to all HVAC units in the ILC control group either in real-time (e.g., RTP or CPP) or beforehand (e.g., TOU prices for winter and summer seasons) while each customer can set up his own response curve by inputting values for $\left[T^{min}_{setpoint}, T^{max}_{setpoint}, B^{min}, B^{max}\right]$ based on his willingness trading comfort for money. In the ILC case, appliances will not report to the central controller what their preferences are. This is because that the response curve of a customer may change frequently in a day. Thus, compared with the DLC-based case, it is more complicated and less accurate for the ILC-based central controller to derive the relationship between electricity tariffs and the aggregated power consumption. This is because the aggregated response curve is highly nonlinear. In addition, the power consumption of the aggregated load may oscillate because the load diversity may be lost without coordination between devices.

3.2.2 *ILC-Based Demand Response under TOU Price*

As shown in Fig. 3.19, a typical TOU rate program has three energy charges: on-peak, shoulder peak, and off-peak. In winter, the peak load occurs between 6:00 and 9:00 a.m. when people wake up and building schedules are adjusted from night-time modes to the day-time modes. Normally, it is also the coldest time of the day, so the heating load will be at its peak during early morning hours. In this example, there are two shoulder load periods: 9:00 a.m. to 12:00 p.m. and 17:00 to 20:00 p.m.

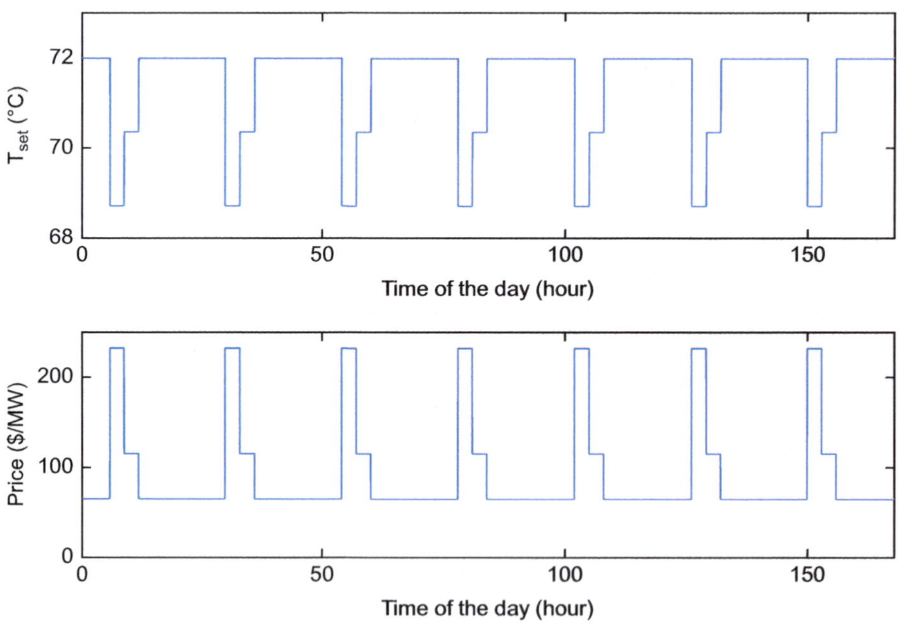

Fig. 3.19 Residential TOU prices for the heating season and the corresponding response curve of an HVAC unit

Let B_{OnPeak}, $B_{Shoulder}$, and $B_{OffPeak}$ represent on-peak, shoulder, and off-peak electricity price, the response curve of an HVAC unit under the TOU price can be represented by

$$T_{set} = \begin{cases} T_{setpoint}^{max} & B = B_{OffPeak} \\ \left(T_{setpoint}^{min} + T_{setpoint}^{max}\right)/2 & B = B_{Shoulder} \\ T_{setpoint}^{min} & B = B_{OnPeak} \end{cases} \tag{3.5}$$

Assume that an ILC-based DR program consists of 1000 HVAC units in their heating modes. Let P_{rated} randomly distributed between 4 and 6 kW and the RCQ parameters of the 1000 units randomized. We will run the simulation for one winter week at 1-min simulation interval.

In the Base Case, the $T_{setpoint}$ is 75 °F and the $T_{deadband}$ is 2 °F. As shown in Fig. 3.20a, room temperatures of all 1000 HVAC units are held between 74 and 76 °F all the time. As shown in Fig. 3.20b, the total power consumption of the 1000 HVAC units is smooth. When the outdoor temperature is above 60 °F, the cycling of the HVAC units is less frequent, causing the total energy consumption of the HVAC control group to drop close to zero.

In the TOU case, set $T_{setpoint}^{max}$ at 75 °F for all 1000 HVAC units and let $T_{setpoint}^{min}$ randomly distributed between 71 and 73 °F. Then, based on Eq. (3.5), a response curve is obtained for each HVAC unit. The resultant room temperature profiles and the total energy consumption in the TOU case are shown in Fig. 3.21a,b. A comparison of the total energy consumption between the no-controlled base case and the TOU case is shown in Fig. 3.22.

From this example, we can see that when responding to TOU signals, the setpoint of the HVAC unit will decrease so that the overall energy consumption in the targeting peak period will decrease, as shown in Fig. 3.22. However, this does not lead to a peak power reduction. Instead, the peak power will increase drastically.

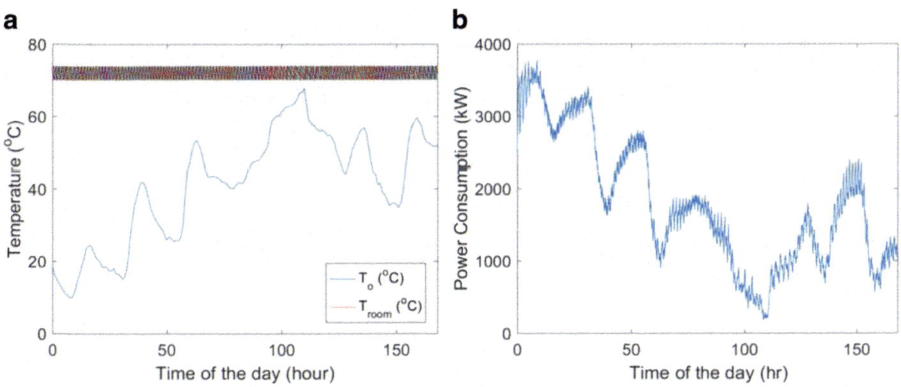

Fig. 3.20 (**a**) Room temperature profiles of 1000 HVAC units and (**b**) total power consumption of the 1000 HVAC units

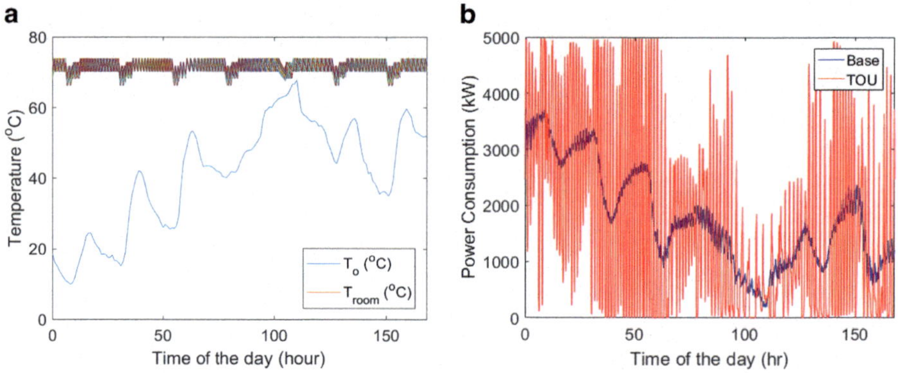

Fig. 3.21 (**a**) Room temperature profiles of 1000 HVAC units and (**b**) comparison of the total power consumption between the base case and the TOU case

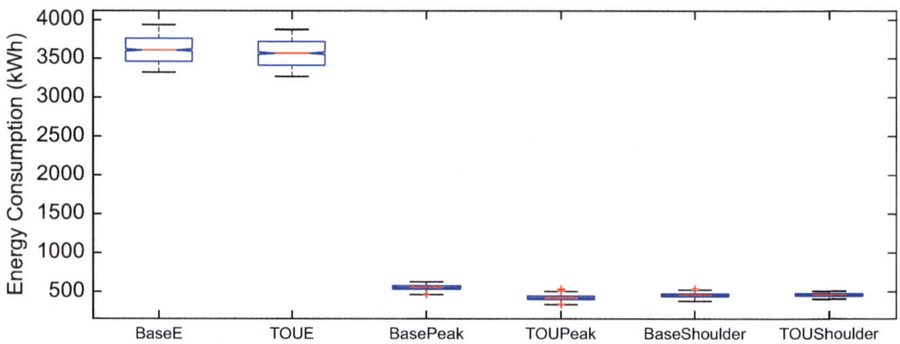

Fig. 3.22 Comparison of total weekly energy, peak-period energy, and shoulder-period energy consumptions when 1000 HVAC units respond to the TOU rate

This is because the cycling of 1000 HVAC units are synchronized after responding to the TOU price signal, as shown in Fig. 3.21b. In addition, energy consumption of HVAC units during the shoulder periods will increase slightly. This is because the setpoint of an HVAC unit will increase when the energy price decreases. Then, the compressor unit needs to be turned on for a longer duration to heat up the room in order to reach the new setpoint. Therefore, the total energy consumption over the shoulder period may increase slightly although the setpoint $\left(T_{setpoint}^{min} + T_{setpoint}^{max}\right)/2$ is still lower than that of the base case (i.e., $T_{setpoint}^{max}$), as shown in Fig. 3.22.

This simple example shows the limitation of using TOU prices for peak load reduction or for shifting energy consumptions from high to low price periods. Thus, for large-scale DR response programs using the ILC-based approach, coordination is necessary to limit the number of HVAC units on the same phase or from the same feeder to preserve load diversity.

3.2.3 ILC-Based Demand Response under the Critical-Peak Price

CPP is designed to reduce the magnitude of the coincident load peak. Let the critical peak price, B_{CPP}, be issued between 6 a.m. and 9 a.m. every day. We will use an HVAC control group of 1000 HVAC units to illustrate the design of a CPP-based DR program. Each HVAC unit will be controlled by a response curve defined by eq. (3.4). To simplify the simulation, set $B_{CPP}^{min} = 20$ \$/MW and $T_{setpoint}^{max} = 75$ F for all HVAC units. To randomize the setting of the HVAC units, let B_{CPP}^{max} and $T_{setpoint}^{min}$ vary between [50 80] \$/MW and [71 73] °F, respectively.

Figure 3.23 shows the response curve of an HVAC unit. When the CPP price is high, $T_{setpoint}$ will be lowered to reduce the average energy consumption over the CPP period.

As show in Fig. 3.24a, the thermostat setpoints of the 1000 HVAC units can drop 2–4 °F from the original setting of 75 °F in response to CPP. Total power consumptions of the 1000 HVAC units compared with those of the base case are shown in Fig. 3.24b. Total energy consumptions of the 1000 HVAC units compared with those of the base case are shown in Fig. 3.25.

From the results, we have the following observations:

- Energy consumptions in the CPP period drop slightly but the energy consumption in the shoulder and low price periods will increase. Overall, the weekly energy consumption of HVAC units will increase.
- Power consumptions of the aggregated HVAC load will increase compared with the base case.

Thus, if the operation of the HVAC units is not well coordinated, it is not guaranteed that the energy consumption of the aggregated HVAC load will be reduced. On the other hand, the maximum power consumption will increase. This is because when each HVAC unit is controlled by its own thermostat, the "on" and "off" of an HVAC unit is random to the whole group. However, when responding to

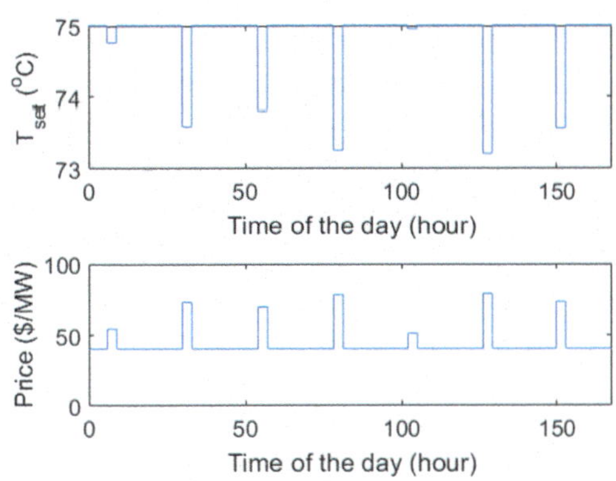

Fig. 3.23 Setpoint changes with respect to critical peak prices for one HVAC unit

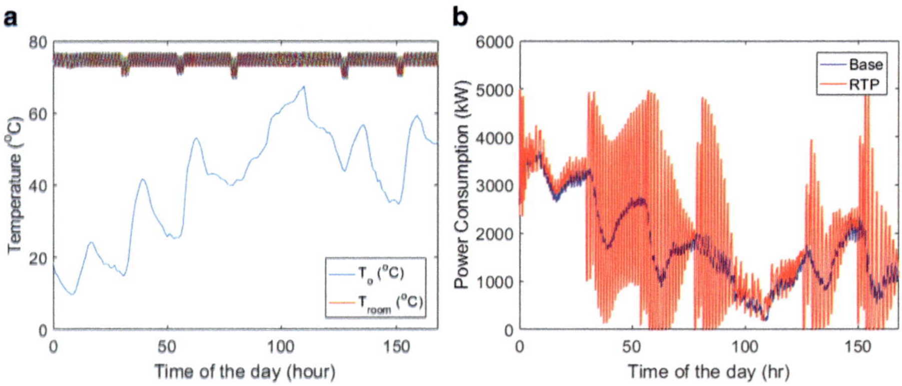

Fig. 3.24 (**a**) Room temperature profiles of 1000 HVAC units; (**b**) Total power consumption of the 1000 HVAC units

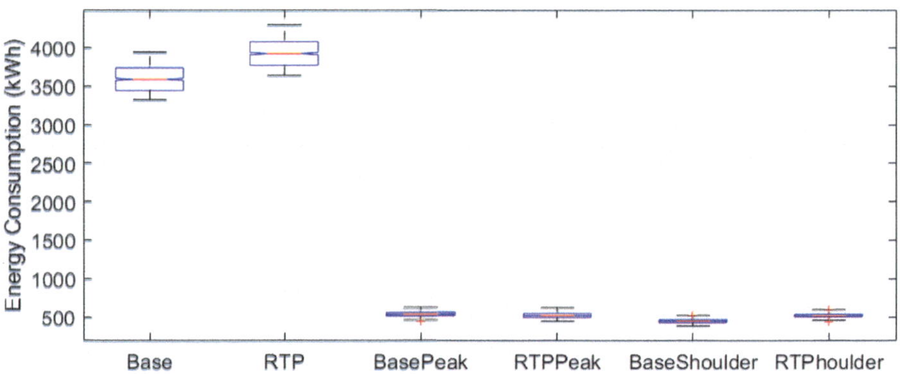

Fig. 3.25 Comparison of total weekly energy, peak-period energy, and shoulder-period energy consumptions when 1000 HVAC units respond to the real-time rate

CPP or TOU prices, the HVAC units will be turned on/off responding towards a common price signal, causing their on/off cycles to be synchronized.

In addition, the HVAC unit will remain "on" or "off" for a longer duration when its thermostat setpoint shifts. Longer "on" durations will cause more HVAC units to be on simultaneously. This will in turn lead to a higher load peaks and lower load valleys, causing large power swings (see Fig. 3.24b).

Those factors must be carefully considered when designing the ILC-based DR program. Otherwise, the following phenomena may occur:

- There are more large power oscillations.
- There are more power peaks.
- Energy consumptions are higher than that of the no-control case.
- Load peaks occur in traditionally nonpeak periods.
- Load peaks have longer durations.

3.3 Autonomous Load Control Algorithm

We will now introduce design considerations of the ALC control approach by considering using ALC for frequency regulation. Traditionally, frequency regulation is provided mostly by generators. Load-side resources are used for Under Frequency Load Shedding (UFLS) as a fast remedial action under emergency conditions. The purpose of UFLS is to balance generation and load when an event causes the system frequency falls below a certain threshold. During such events, under-frequency relays are triggered to shed loads on entire distribution feeders to restore the load-to-generator balance. Thus, the load shedding is nonselective.

Table 3.5 published at the North American Electric Reliability Corporation (NERC) website (https://www.nerc.com/pa/RAPA/ri/Pages/UnderFrequencyLoadShedding.aspx) illustrates a history of UFLS events from 2003 to 2012. As shown in the table, single events had a load shedding range from 24 MW to 17,644 MW. Because of the large range of load lost in UFLS events, NERC is considering to establish a UFLS total load loss threshold. Among the performance metrics, the significance of a UFLS activation compared to the total activation load loss is assessed to understand whether the desired amount of loads had been shed.

Using autonomously controlled, frequency-responsive (ACFR) loads for under-frequency load shedding makes the UFLS selective. Once an ACFR load detects that system frequency drops below a triggering frequency, it can be turned off automatically or switched to a low power consumption mode. Refrigerators, air conditioners, space heating units, water heaters, freezers, dish washers, clothes washers, and dryers are all potential ACFR resources. We are particularly interested in using smart appliances with microcontrollers installed for providing the UFLS service because it will be relatively cheap to use existing sensors and processors for retrofitting an appliance to an ACFR load.

To make a load ACFR, we need to consider the following parameters:

- Triggering frequency, f_{off}, and restoring frequency, f_{on}. One needs to know at which frequency an ACFR load should be curtailed and restored.

Table 3.5 History of UFLS events from 2003 to 2012

Year	FRCC	MRO	NPCC	RFC	SPP	SERC	TRE	WECC
2003			17644	6105			1549	
2004			24					
2005								
2006								
2007		480						
2008	1273		48		673			
2009								
2010								
2011				176				

- Triggering delay, T_t. One needs to know what delays, if any, the ACFR load should observe prior to curtailing or resuming its load?
- Reset delay, T_r. One needs to know how long the ACFR load should wait to resume its operation?

As a load service provider, the following system level questions need to be addressed:

- Should the ACFR loads be distributed all over the system, or should their locations be carefully chosen?
- Because the response is autonomous, is there a need to verify the action of the ACFR load?
- Because each ACFR load acts on its own, how can one make sure the aggregated response is sufficient at any given time?

While asserting and implementing frequency setpoints and delays are not challenging, the ramifications of these design decisions are great. First, the choice of frequency setpoint determines whether the ACFR load will contribute toward small signal or emergency control. The frequency setpoints and delays also define the frequency and durations of the under-frequency events that are recognized and responded to by the ACFR load, and this has a direct impact on the potential inconvenience experienced by an appliance's owner. We also recognize that should the ACFR load become common, their simultaneous actuation could itself create instabilities. Furthermore, long actuation delays might exacerbate dynamic power swings that sometimes accompany grid emergencies.

Indeed, a more clever design of the ACFR mechanism could perhaps attenuate such power swings. First, actuation delays must be fast enough to reduce the initial power swing. Simultaneous actuations can be prevented by designing likelihood, or probability, functions into the ACFR-based frequency responses. Alternatively, the response frequency itself might be randomized about a desired frequency setpoint. The resumption of load is more easily randomized through randomization of turn-on time delays. These issues, while of great interest to smart appliance design, will all need to be addressed in the design of autonomously controlled ACFR load for frequency regulation.

Thus, in this section, we will first examine the characteristics of frequency fluctuations in a given regional power grid. Then, we will introduce control parameters of the ACFR load and operational statistics when varying frequency setpoints and delay parameters. A few fundamental analyses for determining the frequency setpoint and delay design decisions will be presented to illustrate the design considerations and procedure.

3.3.1 *Statistics of Raw Frequency Data*

To understand how often an ACFR load would have been called to service, we will first take a look at the raw frequency data collected from an actual system. In

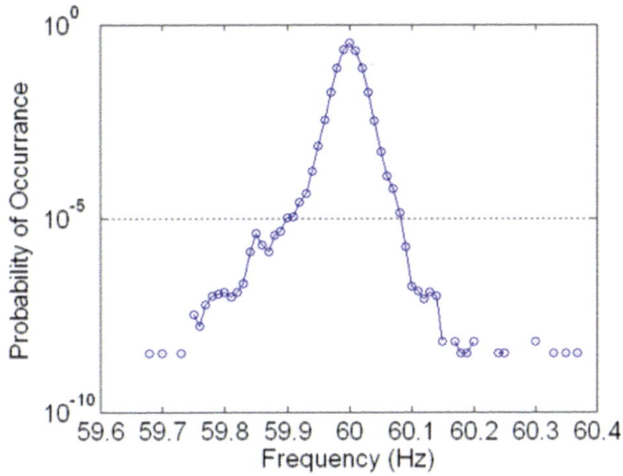

Fig. 3.26 A logarithmic plot of the probability frequency distribution of raw electrical power frequency data from 2003 in the range 59.80 to 60.10 Hz

Fig. 3.26, a logarithmic probability distribution histogram of frequency is presented. The duration of the data is 1 year and the sampling frequency is 10 Hz. All raw frequency data samples are grouped into 30 bins from 59.80 to 60.10 Hz with a bin size of 0.01 Hz. Thus, the sampling duration, $t_s = 0.1\text{s}$.

There are several observations based on the obtained frequency statistics in Fig. 3.26:

- Because the objective of power system operators is to maintain the system frequency very close to 60 Hz, the system frequency follows a normal distribution with a mean at 60 Hz and remains within a narrow range of 59.95–60.05 Hz 99.95% of the time.
- Between 59.81 and 59.95 Hz, except for a peak near 59.85 Hz, the distribution of frequency nearly follows an exponential decay trend.
- The distribution of frequencies under 59.90 has a peak around 59.85 Hz. This may suggest that for many frequency excursions beyond 59.95 Hz, minimums were around 59.85 Hz.

Assuming an ACFR load is placed in this system to provide the UFLS function, we will describe its operation in detail in the next subsection.

3.3.2 Frequency-Responsive Appliance Design Considerations

Define the control parameters that contribute to the control logic design of the ACFR load as follows:

- Frequency setpoints: the turn off setpoint, f_{off}, and the turn on setpoint, f_{on}.

- Control delays: the triggering delay, t_t, and the reset delay, t_r.

We will now explain why each parameter shall be considered and how it is simulated.

The most important design parameters are the two frequency setpoints. One must first determine the control objective. If an ACFR load is set to respond to frequency signals of 60 ± 0.05 Hz, it competes with generators in providing small signal frequency control. If the ACFR load is set to respond to frequency signals lower than 59.95 Hz, it is treated as a spinning reserve and helps remediate larger frequency excursions than can be adequately handled by generator control alone. If the ACFR load is set to respond to frequency signals lower than 59.2 Hz, it will compete with existing substation under-frequency relays and will actuate only after many substation under-frequency relays have already reacted.

Note that f_{on} and f_{off} can be set at different frequency values to provide a dead band and prevent the ACFR load from oscillatory behaviors. The triggering delay t_t can be used to coordinate with generator regulation and avoid reacting to high frequency disturbances.

Raw frequency data sampled at 10 Hz contain a lot of high-frequency noise, as shown by the gray line in Fig. 3.27. To filter out the noise, a filter can be installed on the frequency transducer, as shown by the solid lines in Fig. 3.27.

If we consider the basic frequency curve to be the one obtained by using an average filter with a 20-point moving window, then a maximum filter with a 20-point moving window provides an upper envelope of this frequency curve and results in a delay of 2 s when tracking frequency declines (Fig. 3.28a). The 2 s delay t_t is calculated by.

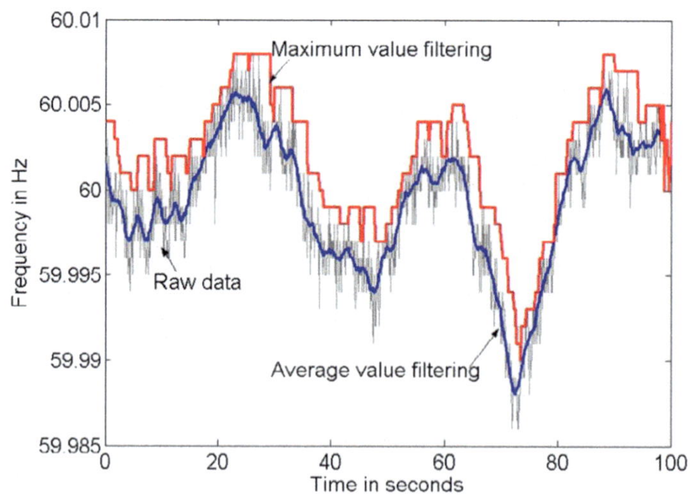

Fig. 3.27 A moving average filter and a moving maximum filter

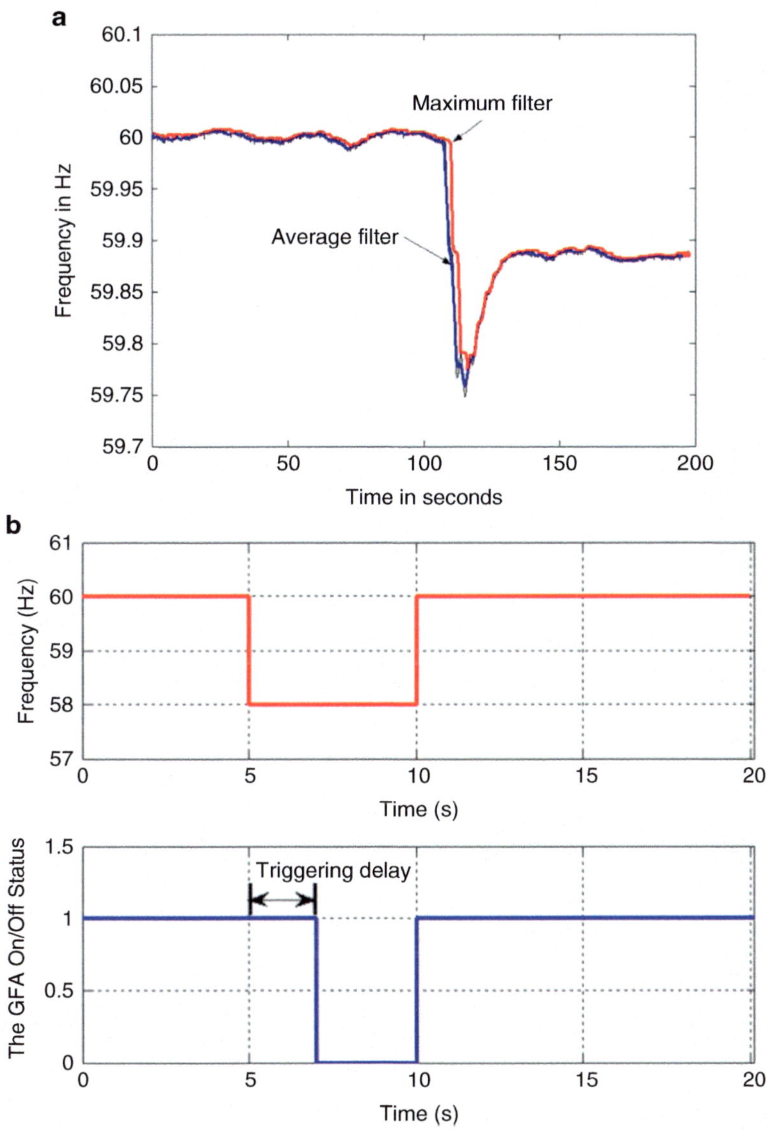

Fig. 3.28 (**a**) A frequency dip detected by using a moving average filter and (**b**) a moving maximum filter, respectively

$$t_t = nt_s = 20 \times 0.1 = 2\text{s} \tag{3.6}$$

where n is the length of the moving window and t_s is the sampling period. Because of the delay, the duration of the detected frequency dip will be shorter than that of the original one, as illustrated in Fig. 3.28b.

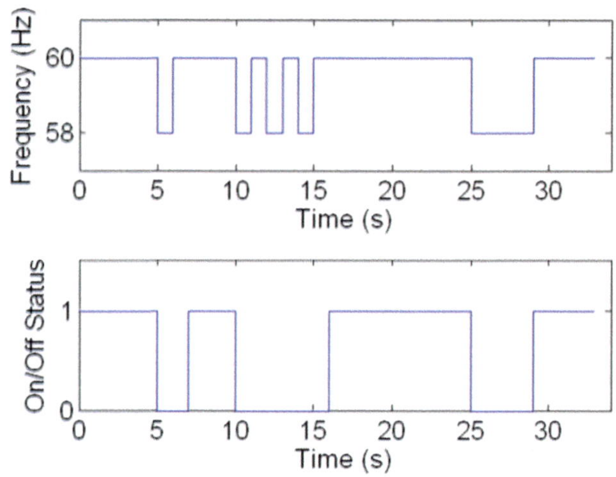

Fig. 3.29 The impact of the reset delay t_r on ACFR for frequency dips

ACFR loads are primarily consumer home appliances. The allowed number of triggering events can be unlimited or limited based on the specific appliance. For example, water heater loads are resistive and can be turned on or off almost as often as needed without inconveniencing the appliance owner. However, heating, ventilation, and air conditioning (HVAC) units can only be turned on or off periodically. Therefore, t_r may be set to provide a minimum cycle interval. One of the principal tenets of "Grid Friendly" appliances is that the appliance manufacturer should be allowed to custom design a response for his appliance that is acceptable for both the performance of the appliance and the consumer who purchases and uses the appliance.

Setting reset delay will force the ACFR load to stay "off" for a designated length of time. As shown in the first subplot in Fig. 3.29, the first and second frequency dips sustain less than the reset delay (2 s in this case); therefore, the ACFR load will respond only once, as shown in the second subplot in Fig. 3.29. For the third frequency dip, which is longer than the 2-s reset delay, the ACFR load response follows the frequency change.

In the next section, the ACFR load triggering frequency and duration parameters are studied using the frequency data shown in Fig. 3.26.

3.3.3 Triggering Statistics of an ACFR Load

Assuming that an ACFR load is controlled by the parameters discussed in Sect. 3.3.2: f_{off}, f_{on}, t_t, and t_r, *we will study the following* three cases: Case 3.1 with no control delays, Case 3.2 with triggering delay, and Case 3.3 with reset delay.

Case 3.1: No Control Delays In the base case, set $f_{off} = f_{on} = f_{set}$ and make f_{set} range from 59.90 to 59.99 Hz. Set triggering delay and the reset delay to be zero. In summary, we have

$$f_{off} = f_{on} = f_{set}, \quad t_t = t_r = 0. \tag{3.7}$$

The base case simulates an ACFR load set to turn off below a frequency threshold f_{set} and turn on above it. We will predict how often the ACFR load will turn off and how long it will stay off when responding to the 1-year frequency data.

Note that we ignore the natural cycling time of the ACFR load. In doing so, we assume that the ACFR load is controlled only by the frequency signal such that when the system frequency is above its setting, it is on; otherwise, it is off. This may not be true in reality. For example, a water heater unit may turn on 1/6 of the time per day. When it is in the "off" state of its natural cycle, a low frequency signal will not turn it off again. Therefore, availability factors will need to be considered without this assumption. Also note that the simulation is an open loop one to obtain basic operational statistics of an individual ACFR load. To simulate the dynamic response of accumulative ACFR loads, one must close the loop of frequency feedbacks. The interested reader is referred to "Grid Friendly™ Device Model Development and Simulation," a report prepared by Pacific Northwest National Lab listed in [7]. In this report, the aggregated impact of ACFR loads on the frequency recovery process is modeled using a dynamic model considering the generator side response as well.

The results of Case 3.1 are summarized in Table 3.6 and Fig. 3.30, from which we make the following observations:

- If an ACFR device responds to a frequency above 59.97 Hz, it will be triggered every 2 min or less, on average. As shown in Fig. 3.8, the frequency at which the ACFR load triggered is greater and the duration for which the ACFR load remained off is much longer than for those ACFR loads having frequency settings lower than 59.97 Hz. Therefore, an ACFR device having a frequency setpoint near 60 Hz should be able to turn on and off frequently and may remain off for extended periods.

Table 3.6 Case 3.1 daily triggering frequency (events/day) in 345 days

f_{set} (HZ)	Number or triggering in a year	Daily triggering frequency	Yearly triggering duration (s)	Daily triggering duration (s)
59.91	177	0.51	1016.4	2.95
59.92	403	1.17	1512.4	4.38
59.93	663	1.92	2440.1	7.07
59.94	2449	7.10	5064.1	14.68
59.95	9946	28.83	15,290	44.32
59.96	48,115	139.46	61,733	178.94
59.97	247,980	718.78	308,020	892.81
59.98	1,115,000	3231.88	1,481,600	4294.49

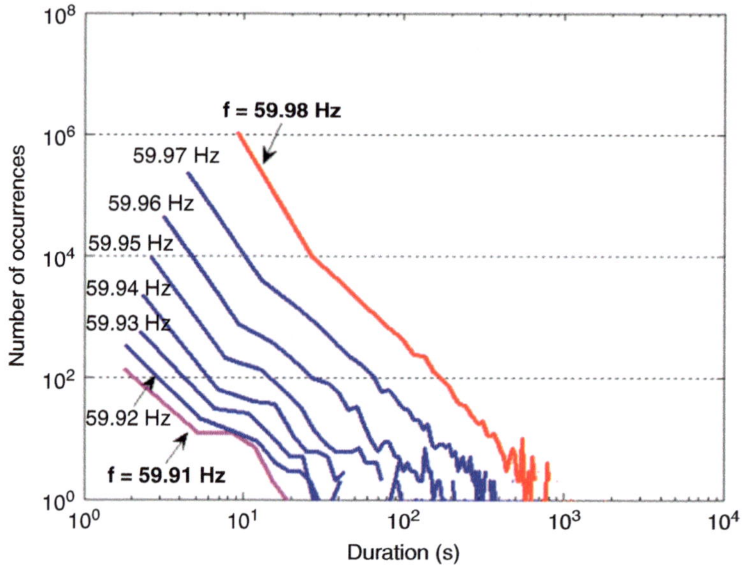

Fig. 3.30 The number of occurrences corresponding to different ACFR loads triggering durations when f_{set} ranges from 59.90 to 59.98 Hz with an increase of 0.01 Hz

- If an ACFR device responds to a frequency ranging from 59.90 to 59.96 Hz, it will be triggered every 10 min or longer, on average. The duration of each event, however, is usually less than 2 s.
- Figure 3.30 also shows that greater frequency disturbances ($f < 59.95$ Hz) are usually short-lived. Most of them are less than 1 s because of the frequency regulation services provided by generators in power systems.

Case 3.2: Triggering Delay In this case, the triggering delay is taken into consideration by using a maximum filter with a moving window of 0.2, 1, and 4 s. We set $f_{off} = f_{on} = f_{set}$ and make f_{set} range from 59.90 to 59.99 Hz. The reset delay is set to be zero. In summary, we have

$$f_{off} = f_{on} = f_{set}, T_t = [0.2\ s, 1\ s, 4\ s], T_r = 0s \tag{3.8}$$

The daily triggering frequency at each frequency setpoint and with different triggering delay is shown in Table 3.7. From the results, we can see that incorporating a triggering delay will make the ACFR load trigger less frequently, and event frequency decreases with increasing trigger delay. This effect is shown dramatically in Fig. 3.31. 59.9 Hz is used as the triggering setpoint f_{set} throughout the three subplots of Fig. 3.31.

The events causing a frequency dip less than 0.2 s will not trigger an ACFR load that is set with 0.2-s triggering delay. The events causing a frequency dip of duration less than 1 s will not trigger an ACFR load that has a 1-s triggering delay, and so

Table 3.7 Case 3.2 daily triggering frequency (events/day) in 345 days

t_t (s) / f_{off} (Hz)	*0*	*0.2*	*1*	*4*
59.9	0.60	0.36	0.12	0.07
59.91	0.51	0.33	0.15	0.10
59.92	1.17	0.73	0.26	0.17
59.93	1.92	1.11	0.39	0.26
59.94	7.10	4.00	1.12	0.56
59.95	28.83	15.59	3.70	1.76
59.96	139.46	73.93	16.21	6.63
59.97	718.78	383.01	84.63	35.02
59.98	3232.01	1749.01	388.63	160.44
59.99	9924.41	5536.25	1298.48	561.10

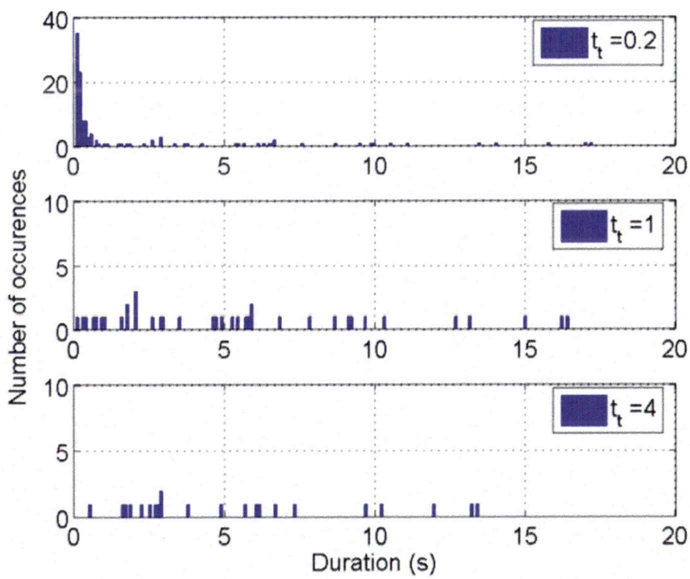

Fig. 3.31 The duration of the ACFR when t_t is 0.2, 1, and 4 s, and f_{set} is 59.9 Hz

on. The second and third subplots in Fig. 3.31 confirm that the number of occurrences with 1- and 4-s triggering delays will be the same as those for no triggering delay with the event durations shifted by 1 s and 4 s, respectively. Those events to the left of the shift are not recognized. This triggering delay effect can greatly reduce the ACFR load triggering frequency because most events are short-lived.

Table 3.8 Case 3.3 daily triggering frequency (events/day) in 345 days

	$t_t = 0.2$ s			$t_t = 1$ s			$t_t = 4$ s		
f (Hz)	$t_r = 1$ s	$t_r = 10$s	$t_r = 100$ s	$t_r = 1$ s	$t_r = 10$s	$t_r = 100$	$t_r = 1$ s	$t_r = 10$s	$t_r = 100$ s
59.9	0.03	0.01	0.01	0.03	0.01	0.01	0.02	0.00	0.00
59.91	0.06	0.01	0.01	0.05	0.01	0.01	0.03	0.01	0.01
59.92	0.09	0.02	0.01	0.08	0.01	0.01	0.06	0.01	0.01
59.93	0.26	0.04	0.02	0.21	0.03	0.01	0.14	0.03	0.01
59.94	0.69	0.10	0.03	0.61	0.10	0.03	0.51	0.07	0.02
59.95	1.98	0.34	0.11	1.59	0.26	0.09	1.22	0.20	0.06
59.96	9.97	1.67	0.54	8.14	1.28	0.41	6.25	0.99	0.30
59.97	55.08	8.95	257	44.44	7.10	2.08	33.78	5.36	1.64
59.98	269.77	41.98	10.13	224.74	34.56	8.44	177.13	26.94	6.84
59.99	1009.87	148.10	2815	871.40	126.57	24.94	722.03	103.72	21.44

Thus, one can use a triggering delay setting on those ACFR loads that cannot be cycled on or off often to make them respond only to longer duration frequency disturbances.

Case 3.3: Reset Delay In this case, a reset delay is also taken into consideration. Set $f_{off} = f_{on} = f_{set}$ and make f_{set} range from 59.90 to 59.99 Hz. Both the triggering delay t_t and reset delay t_r are allowed to vary. In summary, we have

$$f_{off} = f_{on} = f_{set}, t_t = [0.2\ s, 1\ s, 4\ s], t_t = [1\ s, 10\ s, 100\ s]. \tag{3.9}$$

From the results shown in Table 3.8, we noticed that

- The reset delay will also significantly decrease the number of triggered ACFR events, because most of the frequency deviation beyond 59.95 Hz is short-lived.
- A reset delay will prevent the ACFR device from responding to the frequency oscillations, as shown in Fig. 3.32.
- The function of a reset delay is a little different from that of the triggering delay. If randomized between appliances, the reset delay can be used to provide load diversity while ACFR loads begin releasing control of their appliance loads and can prevent the ACFR loads from turning on simultaneously.

3.4 Summary

In restructured power systems, load response programs will add more flexibility in power system control. Traditionally, interruptible loads have been used as the last resort to provide fast frequency recovery when the system frequency drops below normal operation range. Nowadays, load can provide frequency regulation in power system normal operation states and compete with generators in the reserve market.

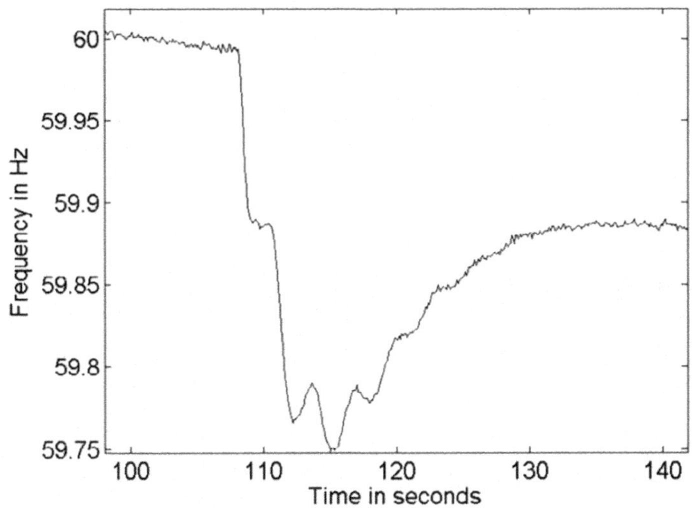

Fig. 3.32 A severe under-frequency disturbance

In this section, we only present a preliminary study on the control design of frequency responsive appliances. Through the rudimentary analysis of under-frequency control parameters based on measured electrical power system frequency data, we hope to give our readers an idea what to expect if an ALC device is deployed. A more complicated scenario will be to study the ramifications of applying the specific control approaches suggested here. Simulating the aggregated ACFR load dynamic responses and modeling the coordination of the frequency regulation service provided by the load and the generator will then be necessary.

References

1. Vanouni, M., & Lu, N. (2016). A resource prioritization based framework for participation of thermostatically controlled loads in regulation services. *IEEE Transactions on Smart Grid.*
2. Lu, N., & Zhang, Y. (2013). Design considerations of a centralized load controller using thermostatically controlled appliances for continuous regulation reserves. *IEEE Transactions on Smart Grid, 4*(2), 914–921.
3. Lu, N. (2012). An evaluation of the HVAC load potential for providing load balancing service. *IEEE Transactions on Smart Grid, 3*(3), 1263–1270.
4. Zhang, Y., & Lu, N. (2013). Parameter selection for a centralized thermostatically controlled appliances load controller used for intra-hour load balancing. *IEEE Transactions on Smart Grid, 4*(4), 476–486.
5. Zheng, L., Lu, N., & Cai, L. (2013). Reliable wireless communication networks for demand response. *IEEE Transactions on Smart Grid, 4*(1), 133–140.
6. Vanouni, M., & Lu, N. (2015). Improving the centralized control of thermostatically controlled appliances by obtaining the right information. *IEEE Transactions on Smart Grid, 6*(2), 108–118.

7. Lu, N., Hammerstrom, D., & Patrick, S. (2009). Grid FriendlyTM device model development and simulation. PNNL-18998. In *Pacific northwest National Laboratory*. Washington: Richland. Retrieved from https://www.pnnl.gov/main/publications/external/technical_reports/PNNL-18998.pdf.

Chapter 4
Demand Responses in ERCOT

4.1 Introduction

The Federal Energy Regulatory Commission (FERC) defines demand response as:

> Changes in electric usage by end-use customers from their normal consumption patterns in response to changes in the price of electricity over time, or to incentive payments designed to induce lower electricity use at times of high wholesale market prices or when system reliability is jeopardized.

The end-user customers who participate in demand response may be very diverse, ranging from small-size residential load to large commercial and industry load. These customers are mainly attracted to the demand response programs for a financial payment in an exchange for a change to their energy usage either during emergencies or normal situations. The changes to normal usage patterns can incur as a result of shutting off, not starting, or reducing power consumption in power consuming equipment; starting on-site generation or relying on alternate fuel sources if they are available. These characteristics and capabilities of end-user customers also drive the design and development of a broad range of demand response programs with different performance requirements and compensation mechanisms in which loads participate.

Essentially, demand responses differ significantly in when and how these actions could take place. These differences constitute the rationales for how demand responses are categorized. These differences include, but do not limit to, the reasons for the demand response action (signal or trigger), the amount of time given in advance of the required action (ramp period), the length of time the demand response event lasts (sustained response period), and measurement and verification of the response.

According to the North American Electric Reliability Council (NERC), demand response can be broken down into two major categories: dispatchable or non-dispatchable. The former are those demand response events that are initiated by utilities and the latter are defined as an event not specifically initiated by utilities.

P. Du et al., *Demand Response in Smart Grids*,
https://doi.org/10.1007/978-3-030-19769-8_4

4.2 Overview of ERCOT

The Electric Reliability Council of Texas (ERCOT), a nonprofit corporation, is the Independent System Operator (ISO) formed to ensure reliability of the electricity grid. The ERCOT region covers about 85% of Texas, including all areas that are open to retail competitive choice. As one of ten North American ISOs, ERCOT's primary responsibility is to maintain the reliability of the electric grid by ensuring a precise balance between load and generation throughout its service area on a second-by-second basis, using market-based mechanisms to the fullest extent possible. ERCOT administers and facilitates market rules which are developed in a collaborative environment with stakeholders and regulators.

The Public Utility Commission of Texas (PUCT) has established a goal to "ensure not only that the load resources that have historically participated in the markets have reasonable opportunities to continue to participate, but also that a framework is established for even greater participation by load resources in the future." In addition, the Texas Legislature in 2011 passed Senate Bill 1125, which establishes new goals and requirements for energy efficiency and also requires the PUCT to adopt rules and procedures "ensuring that ERCOT allows load participation in all energy markets for residential, commercial and industrial classes, either directly or through aggregators of retail customers, to the extent that load participation by each of those customer classes complies with reasonable requirements adopted by the organization relating to the reliability and adequacy of the regional electric network and in a manner that will increase market efficiency, competition, and customer benefits."

Accordingly, demand response in the ERCOT Nodal market is viewed as a means of enhancing competition, mitigating price spikes, encouraging the demand side of the market to respond better to wholesale price signals, providing resource adequacy, and preserving system reliability.

4.3 ERCOT Market Structure

The primary role of ERCOT is to maintain reliability of the ERCOT electricity grid, using market-based mechanisms to the fullest extent possible. As the Independent System Operator (ISO), ERCOT is charged with maintaining a precise balance between load and generation on a second-by-second basis.

ERCOT administers markets to assist market participants in buying and/or selling capacity and energy to meet their energy and ancillary services requirements. To facilitate this, ERCOT administers various Day Ahead and Real-Time operations markets where buyers and sellers of capacity and energy may participate.

Wholesale electricity market prices fluctuate based on simple rules of supply and demand—that is, the relationship between the level of power consumption (demand) being placed on the grid at any time and the generation resources (supply) available to meet that demand. During periods of peak demand, such as hot summer days, wholesale prices may reach many multiples of their off-peak levels. Even during milder weather when demand is lower, premium prices are often paid for available resources because many generation plants choose these times to shut down temporarily for maintenance.

The ERCOT Nodal market is designed with a number of features to reward energy consumers that are willing to curtail or shift load as a way of helping maintain system reliability. These "demand-side resources," or loads, are encouraged to make their resources available by responding to wholesale price signals. Actual dollar values to be paid for these resources are established in the form of Market Clearing Prices, which in a Nodal market are based on bids and offers in the various ERCOT-operated markets described below. The Market Clearing Price for Capacity (MCPC) expressed in dollars per megawatt per hour represents the price paid for making a capacity resource (load reduction or generation increase) available to the ERCOT Grid. The Locational Marginal Price (LMP) expressed in dollars per megawatt-hour represents the price paid for generation at a specific location on the ERCOT Grid. The same calculation engine that ERCOT uses to dispatch generation also calculates the value of energy at every load point in the Transmission System. The individual load-weighted average of all the LMP prices in a Load Zone is the price paid by load serving entities who serve customers with the energy from the generation. This price is sometimes referred to as the Load Zone Market Price or LMPZ. Since this price is calculated in real time, it is not known to energy consumers in advance but rather is posted to ERCOT's website immediately following each calculation of LMPZ. ERCOT has also implemented a system of projecting nonbinding prices, covering the next hour.

4.3.1 Ancillary Services (AS)

Ancillary services are defined in the ERCOT Nodal Protocols as "services necessary to support the transmission of energy to Loads while maintaining reliable operation of the Transmission Service Provider's transmission system using Good Utility Practice." ERCOT requires Retail Electric Providers (REPs) to carry a specified level of operating reserves, which provides the ISO with the ability to call up additional resources on varying levels of short notice. These operating reserves serve as "insurance" in case a generating unit goes down, load is higher than anticipated, or another problem emerges. Such reserves can be self-arranged. If an REP doesn't self-arrange all its reserves, then ERCOT will purchase the necessary reserves on the REP's behalf. Such procurement occurs in a Day Ahead Market for the various Ancillary Services, and may also be procured through a Supplemental Ancillary Services Market (SASM) if a shortage occurs in real time. In the Nodal market, there are eleven AS programs in place, four of which are available for participation by loads.

4.3.2 The Day Ahead Energy Market

The Day Ahead Market (DAM) is an energy market where Qualified scheduling entities (QSEs) may sell or buy energy at any settlement point on the ERCOT Grid. Settlement points exist at each generator's electrical bus (Resource Node) in the transmission system, at trading Hubs and at Load Zones. In the Nodal market, an REP pays ERCOT for all the energy used by its customers, based on a price of energy established every 15 min in ERCOT's Real-Time market and adjusted for any energy purchased in the DAM. REPs may offer load customers the ability to also purchase all or a portion of their energy in the DAM. Similarly, loads may purchase their supply at a specified trading Hub with additional charges applying to the difference in the Hub price and the Load Zone prices. Such arrangements are virtually unlimited and allow loads to purchase their energy needs in these markets or in combination with traditional bilateral contracts or contracts for differences. Loads may choose to place bids in ERCOT markets to purchase energy only "up to" a certain price limit. This practice exposes the load to real-time prices, which may be higher or lower than the limit specified by the purchaser in a specific interval, and effectively puts the Load in the position of bidding to provide demand response if prices clear at or above the specified level.

4.3.3 The Real-Time Market

When a customer uses energy, its supplier must pay ERCOT for that energy at prices established by the Real-Time market, adjusted for any energy purchased in the Day Ahead market at the same load zone. If a load-serving entity does not purchase energy in the Day Ahead market, then it is settled at the Real-Time market price. A customer may participate in the Real-Time market by requesting its REP to provide pricing equal to the Real-Time prices and then adjusting its usage to consume less energy when prices are high (or more energy when prices are low). However, the actual real-time prices that are faced by a consumer exposed to the Real-Time market will not be known by the consumer until after-the-fact, since they are calculated on a real-time basis.

4.4 Overview of Demand Response Options in ERCOT

The demand response programs in ERCOT have been continuously evolving and it has successfully attracted a large number of participants since the inception of the market operations due to several reasons. First, there are a spectrum of options available for the end-user customers to choose from. From those options, a demand-side resource can evaluate its ability to participate in different demand response services: response time, recovery time, predictability of electrical demand, and flexibility of operations, and find out the best program in which its ability can be utilized. Second, the operational and regulatory barriers for load resources to provide demand response services in Texas have been reduced to the great extent. Third, the industry loads, which have more flexibility in their scheduling process, are very active in demand response programs. With an anticipation for a steady source of revenues, those load resources are willing to adjust their energy consumption to improve the grid reliability. Therefore, the size of demand responses in ERCOT has been gradually growing in the past and they are playing an increasingly critical role in ensuring the efficiency and reliability of the ERCOT grid.

These demand responses in ERCOT can fall into two categories: non-dispatchable events and dispatchable demand response.

These non-dispatchable events may include decisions made by the end-use customer to alter its usage pattern or may include contractual obligations with another entity, such as the end-use customer's REP, to alter usage patterns. In ERCOT, non-dispatchable demand response can include a response to Four Coincident Peak (4CP), LSE contracted response to wholesale market energy prices, self-directed response from prices in a retail price contract or Tariff, prices in the Real-Time or Day Ahead markets, or utility-managed Load Management programs (LM).

Several types of dispatchable demand response can participate in the wholesale market, which are deployed for grid reliability. Loads providing Emergency Response Service (ERS) may qualify for a 10-min ramp period or a 30-min ramp period, and either can be provided by a category of resource ("weather-sensitive loads") with DR driven by air conditioning controls. Registered and qualified Load Resources can provide Reserves in either Responsive Reserve Service or Non-Spinning Reserve Service. Regulation in the form of Up-regulation Service and/or Down-Regulation Service can currently be provided by a special type of demand resource called a Controllable Load Resource. In addition, Controllable Load Resources may qualify to submit offers into and be dispatched by Real-Time Energy Market (RTEM).[1]

4.4.1 Non-ERCOT-Dispatched DR

The options listed below are not dispatched by ERCOT but they may be dispatchable by another entity such as an REP or a transmission and distribution utility.

[1]Security Constrained Economic Dispatch (SCED).

4.4.1.1 Four Coincident Peak (4CP)

Interval Data Recorder (IDR)-metered customers are subject to transmission charges based upon a Four Coincident Peak demand. The 4CP demand is determined by averaging the consumer's actual demand during the settlement interval that occurs during the settlement interval of the highest ERCOT demand during each of the four summer months (June through September). This measured 4CP demand serves as the basis of the customer's transmission tariff charges for the following year. By correctly predicting the ERCOT system peaks during the summer and curtailing load during those intervals, a consumer can help reduce the stress imposed on the electric system during peak periods of consumption and reduce their transmission charges for the following calendar year.

4.4.1.2 LSE Contracted Price Response

In the competitive areas of ERCOT, consumers can contract with their REP to respond at the REP's direction. The contract will usually outline the parameters of this response – at what times and frequency demand response events can be called, ramp periods, sustained response periods, compensation, etc. Because this response is a contractual matter between the REP and the consumer, a great deal of variety can be present in these arrangements. For example, a consumer's response may be voluntary or required; compensation could come in the form of reduced energy prices or rebate payments for each curtailment event; consumers might be notified up to a day in advance, or could have no notification at all (for automated curtailment).

In areas of ERCOT not open to competition, interruptible tariffs may be available. These tariffs will usually offer a reduced energy price for defined curtailment obligations.

4.4.1.3 Self-Directed Price Response

REPs may offer dynamic pricing options, or consumers within a municipality or cooperative may have their energy prices determined by a published tariff, which also may be structured based upon Time-of-Use (TOU) or have a Critical Peak Pricing (CPP) or Peak Time Rebate (PTR) component. Self-directed price response refers to consumers making an independent decision to respond to energy prices contained in a governing tariff or in either the ERCOT Day Ahead or Real-Time energy markets.

Consumers within a municipality or cooperative may have their energy prices determined by a published tariff. Tariffs may be structured based upon TOU, CPP, or PTR.

TOU offerings will typically have higher energy prices during traditionally peak periods – for example, a TOU tariff may charge one price during Monday through Friday from 2 to 8 p.m. and another price during all other times. Customers may choose to reduce consumption during these high priced periods. Load reduction can be accomplished by load shifting (loads can be shifted by rescheduling certain processes or by utilizing thermal storage), ending certain processes that are no longer economic, or through energy efficiency measures.

Offerings that incorporate CPP usually have prescribed high prices only during certain defined periods – for example, when load or prices are expected to reach a certain level. Response to CPP may be similar to those employed under TOU tariffs, especially if CPP is reached frequently. More often, CPP is infrequent and as such, short-term load curtailment may be the more appropriate response to meet economic objectives.

4.4.1.4 Transmission and Distribution Utility Commercial Load Management (CLM) Programs

The law which restructured the state's electric utility industry, and PUCT rules require each investor-owned transmission and distribution utility to meet a percentage of their annual growth in system demand through savings achieved by energy efficiency programs. The PUCT has determined that the Transmission and Distribution Utilities (TDUs) may implement and manage Load Management Standard Offer Programs, which pay participating eligible customers for agreeing to provide demand response when needed to help preserve electric grid integrity. Load Management is listed as an approved energy efficiency program to help meet the load growth reduction goals. These programs are not administered by ERCOT; however, the TDUs have executed agreements with the ISO to enable ERCOT to benefit from this demand response capacity during system emergencies.

4.4.2 ERCOT-Dispatched Demand Response

4.4.2.1 Emergency Response Service (ERS)

ERCOT procures ERS to maintain grid stability during emergency conditions and reduce the likelihood of the need for rotating outages. ERS participants may offer to provide demand response with either a 10 min ramp period requirement (similar to Load Resources providing Responsive Reserve Service) or a 30-min ramp period requirement. ERS Resources do not have the same telemetry and under-frequency relay (UFR) requirements as Load Resources providing responsive reserve services.

ERS is defined by the PUC as a special emergency service (not an Ancillary Service). It is procured through an request for proposal (RFP) process three times per year, for 4-month contract terms, each of which is further split into smaller time periods based on business cycles and other factors. Loads may choose to submit offers in all time periods or only in those that best fit their unique circumstances, and may vary both the price and DR capacity offers by time period. Settlements are performed with the QSE up to 80 days after each 4-month period has ended.

Payments are subject to downward revision based upon delivered demand response capacity and the load's availability during the contract term.

ERS is open to aggregations of customers and is explicitly intended to attract customers of all types and sizes. ERS is also open to certain types of generators (those not registered with ERCOT as Generation Resources) that are capable of injecting power onto the ERCOT System. In addition, during the summer contract term (June through September), a different type of ERS Resource—Weather-Sensitive ERS Load—may participate under modified rules designed to accommodate air conditioning load.

4.4.2.2 Ancillary Services Markets

ERCOT is responsible for developing a daily Ancillary Services (AS) Plan that procures AS in sufficient quantities to maintain the security and reliability of the ERCOT System consistent with ERCOT and North American Electric Reliability Corp. (NERC) standards. ERCOT determines the amount of AS that each market participant is required to provide, based on its ratio share of the total ERCOT load. Each market participant may self-provide its AS or may elect to have all or a portion of its AS procured by ERCOT in the Day-Ahead Market. ERCOT will deploy the AS in order to maintain system reliability.

Customers with demand response capability that can meet various performance requirements can be qualified to provide Ancillary Services as *Load Resources*. In the eligible AS markets, the value of having a Load Resource available to reduce load is equal to the value of having a generator available to increase its generation at a generating plant. These providers of operating reserves selected through an ERCOT AS market are eligible for capacity payments, regardless of whether the Resource is actually deployed (or curtailed, in the case of the Load Resource).

Tables 4.1 and 4.2 describe the various Load Resource types and their qualification requirements and eligible services.

Each QSE may arrange bilaterally for AS or may have ERCOT make the arrangements through a market (i.e., an auction) administered by ERCOT. ERCOT holds auctions on a daily basis for each of the following AS:

- Regulation Up.
- Regulation Down.
- Responsive Reserve.
- Non-Spinning Reserve.

Loads that agree to reduce when directed by ERCOT via their QSE, and that meet other metering and operational requirements as specified in the ERCOT Protocols, may participate in AS market auctions described above. As noted in Table 4.1, the type of ancillary service that a Load Resource may provide will depend upon the load's response time and metering system, as well as other requirements described in the protocols. In the Responsive Reserve and Non-Spin Reserve markets, the QSE will receive capacity payments regardless of whether or not the load was actually deployed, but the load must be available for deployment at any time while providing the service.

Table 4.1 Load resource types and ancillary services

Service	Load resource type	Qualification
Responsive reserves (≤60%)	Non-controllable load resource (NCLR)	Under-frequency relay and 10-min response to manual dispatch instruction
Responsive reserves	Controllable load resource (CLR)	Primary frequency response and follow SCED 5-min dispatch
Regulation-up	CLR	Primary frequency response and response to automatic generation control (AGC-type) regulation deployments
Regulation-Down		
Non-spin reserves	CLR	Follow SCED 5-min dispatch

Controllable Load Resource is a Load Resource capable of controllably reducing or increasing consumption under dispatch control by ERCOT

Table 4.2 Requirement for ancillary services

Type of service	Down Reg	Up Reg	Resp Res	Non-spin
System response to instruction:	Must be on AGC	Must be on AGC	AGC or relay action	Response within 30 min
Generation resources	X	X	X	X
Load with under frequency relay installed and capable of being deployed within 10-min notice			X	X
Load with real-time telemetry and that can be deployed within 30-min notice				X
Controllable load resource	X	X	X	X

ERCOT qualifies Load Resources to be able to provide AS based on the load's available technology, as follows:

Responsive Reserve Service (RRS)

RRS requires that an Under Frequency Relay (UFR) be installed that opens the load feeder breaker on automatic detection of an under frequency condition. Alternatively, a Controllable Load Resource must have the capability of responding to frequency change, similar to generator governor response. Both types of Load Resources providing RRS are also required to provide their committed demand response capacity within 10 min following an instruction. Load Resources are randomly assigned to one of two deployment groups for response to such notice. ERCOT may deploy only one of these groups or both of these groups either concurrently or successively depending upon the reliability needs. The load, breaker status, and relay status, if applicable, must have real-time telemetry to ERCOT (through the QSE) installed.

Non-Spinning Reserve Service (Non-Spin)

Non-spin requires that loads provide their committed demand response capacity within a 30-min notice, based on a dispatch instruction. Load Resources providing non-spin must also have real-time telemetry installed, and must be qualified to be dispatched by ERCOT's Real-Time Market software, Security Constrained Economic Dispatch (SCED).

Up-Regulation Service and/or Down-Regulation Service

Regulation service requires that loads through automatic controls respond by changing usage in response to signals provided by ERCOT (similar to Automated Generation Control, or AGC signals) to increase and decrease load while meeting rigorous performance monitoring criteria. Controllable Load Resources (CLRs) must also have real-time telemetry installed and be able to receive certain control information from ERCOT every 4 s. Loads qualified for Regulation Up and Down service are also qualified to provide Responsive Reserves and Non-Spinning Reserves.

4.4.3 Real-Time Energy Market

SCED, the primary dispatch in ERCOT's real-time energy market, is currently open to participation by Controllable Load Resources (CLRs) represented by their Load Serving Entity QSE (LSE QSE). CLRs may be single-site Loads, or may consist of aggregations of smaller Loads such as residential sites with air conditioning compressors under direct load control. SCED is executed every 5 min, so CLRs in SCED with bids at the marginal LMP must be capable of moving load incrementally in either direction every 5 min, based on SCED dispatch instructions (a.k.a., base points). CLRs providing non-spin or RRS are dispatched by SCED after their AS capacity is released to the real-time market.

4.5 Load Resources Providing RRS

Load resources can act as a resource to provide RRS (LaaRs[2]) in the ERCOT market. When the ERCOT market was restructured (Senate Bill 7, effective on Jan. 1, 2002), PUC urged market participants to develop new vehicles for load participation in

[2]**Load Acting as a Resource** (Laa R) is a load that can interrupt in response to a request from ERCOT under various Ancillary Services programs. LaaRs are eligible for capacity payments for making their loads available for curtailment. If they are deployed, they also receive energy payments for actually delivering the load reduction.

RRS. Since then, LaaRs can compete directly with the generation supply and became the market-based replacement for interruptible tariffs.

To provide RRS, LaaRs must have an UFR to be installed that opens the load feeder breaker on automatic detection of an under frequency condition. These loads are also required to be manually interrupted within a 10-min notice. The load, breaker status, and relay status must have real-time telemetry to ERCOT. Table 4.2 details the requirement for load resources in the provision of RRS and compares with other ASs for participation by loads. In 2017, market participants registered an additional 56 Load Resources representing 890 MW of additional capacity based on an aggregation of the maximum interruptible capacity for each Load Resource. At the year's end, there were 351 UFR type Load Resources with an aggregated capacity of 4715 MW that were capable of participating in RRS. It's important to note that at any given time not all of these Load Resources are at their maximum load value, nor are they all participating in the RRS market.

Load Resources controlled by high-set under-frequency relays continue to dominate the number and capacity volume of demand response resources that participate in the ERCOT AS markets. LaaRs also regularly contribute to a large share of the RRS needed. The expected amount of RRS that is required for any given hour is given in the AS Plan requirements.

The details for the methodologies to determine how much RRS is required are provided as follows. ERCOT Operating Guide sets the minimum RRS requirement for all hours under normal conditions. ERCOT will procure amounts of RRS that vary by hour of the day and by month. These RRS amounts will be published by month in six separate blocks covering 4-h intervals. These amounts will be based on expected diurnal load and wind patterns for the month, will cover 70% of historic system inertia conditions for each block of hours for the month, and will use the equivalency ratio for RRS between Load Resources and Generation Resources to establish the conditions for each block of hours. The equivalency ratio will be used to establish the total reserves assuming the Day-Ahead Market (DAM) will use a one-to-one equivalency ratio. The minimum level of RRS procured from Resources providing RRS using Primary Frequency Response shall be determined for each month by ERCOT through the use of studies and shall not be less than 1150 MW. The remaining capacity required for RRS will be procured from all Resources qualified to provide RRS including Load Resources. DAM will limit the RRS procured from Load Resources to 60% of the total RRS requirement. ERCOT may increase the minimum capacity required from Resources providing RRS using Primary Frequency Response if it believes that the current posted quantity will have a negative impact on reliability or if it would require additional Regulation Service to be deployed. RRS amount will be published as a monthly requirement along with the equivalency ratio for each 4-h block. These annually published amounts are the minimum quantity that will be procured in the DAM for each hour of the year.

One type of Responsive Reserve is Interruptible Responsive Reserve. Interruptible Responsive Reserve is provided by Load Resources that are automatically interrupted when system frequency decreases to 59.7 Hz. The amount of RRS procured from these types of Resources during any given hour will be limited to 60% of the total RRS requirement for that hour. The ERCOT Protocols state, "[t]he amount of Resources on high-set under-frequency relays providing RRS will be limited to 60% of the total ERCOT RRS requirement." Self-arranged RRS used to fulfill a QSE's RRS requirement will be limited to 60% from Load Resources excluding Controllable Load Resources.

For 2017, the amounts of ERCOT RRS requirement varied from about 2300 MW on a peak day during the summer to about 2800 MW for an off-peak hour during the fall, winter, and spring months. The awards for UFR type Load Resources were limited to 50% of the total RRS requirement for any given hour.[3] The limit was imposed because of the concern for a frequency over-shooting which could result from the potential nuisance trip of LaaRs when they carry a large portion of RRS responsibility. Figure 4.1 depicts the maximum amount of RRS allowed for LaaRs to participate in 2017. These quantities are also highly correlated with the expected system inertia conditions. In other words, when the system inertia value is higher, the amount of LaaRs allowed to be awarded in the market will be lower.

When cleared in the day-ahead market, LaaRs are paid at the market clearing price of capacity in the RRS market in the same manner as the generation resources do. Under the nodal market, ancillary services and energy are co-optimized in the day-ahead market. Since the bidding prices from generators reflect their marginal cost of reserving additional capacity, these bids are usually higher than those from LaaRs which rank at the low stack of the system bidding curve. Therefore, the market clearing price for RRS is mainly determined by the marginal generator's bid. Because ancillary service clearing prices explicitly account for the opportunity costs of selling energy in the day-ahead market for generators, RRS prices should generally be correlated with day-ahead energy prices. Table 4.3 compares the market clearing price for RRS with other ancillary services. Its value per one unit of MWh is highest among four categories of ancillary services. In 2017, monthly average prices for RRS varied from $7 to $13 per MWh, with the highest price occurring in January. One possible explanation for this decoupling from day-ahead energy prices is that unit commitment patterns have changed because of high wind generation and less online generation capacity capable of providing reserves. This reduction in online capacity, especially in off-peak periods has led to higher prices for RRS in shoulder months. If this decrease in online capacity could be substituted by more LaaRs, the high price would be reduced and a more efficient market operation can be attained.

LaaRs are only deployed either automatically when the system frequency is less than 59.7 Hz or manually by operation instructions to alleviate a generation capacity shortage problem which could lead to an emergency. Overall, the chance for the deployment of LaaRs is very low. Table 4.4 gives the incidents between 2012 and 2016 when LaaRs were taken offline. From this, it can be seen that LaaRs were called upon no more than three times each year between 2012 and 2016. The reason why there were only a few number of deployment of LaaRs is that they are resorted

[3]This maximum limit has been increased from 50 to 60% in 2018.

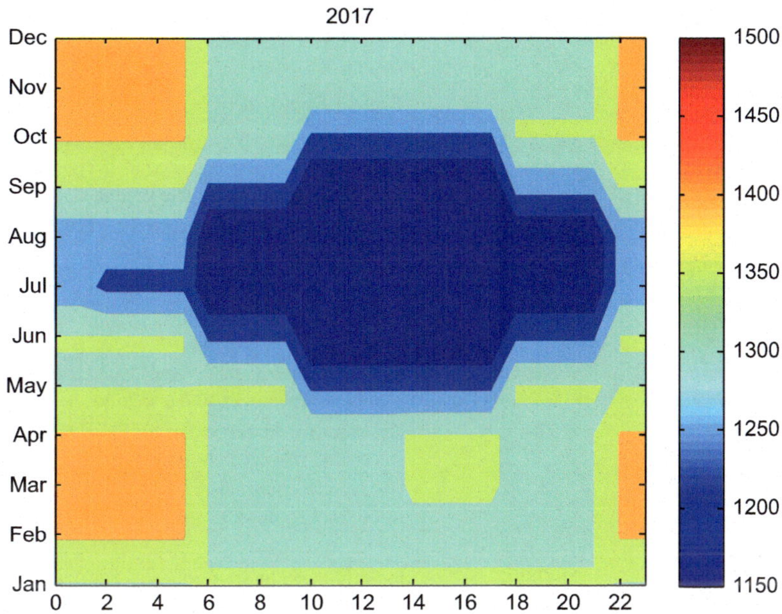

Fig. 4.1 Maximum amount of LaaRs allowed to participate in RRS (MW)

Table 4.3 Average annual ancillary service prices by service

	2016 ($/MWh)	2017 ($/MWh)
Responsive reserve	$11.10	$9.77
Non-spin reserve	$3.91	$3.18
Regulation up	$8.20	$8.76
Regulation down	$6.47	$7.48

as the last safety defense to severe contingencies since LaaRs can effectively respond to the system emergency. This definitely has an impact over how LaaRs are operated. LaaRs need to give a consideration to the duration and frequency of interruption when they design and schedule their operations. It also requires that LaaRs should be able to restore their load consumption within 3 hours once they are tripped so that they can be ready for the next incident. Currently, a majority of LaaRs fall into the category of industrial load.

One example of LaaRs deployment is depicted in Fig. 4.2. During that day, 20% of generation fleet (14,000 MW) were out of service on planned maintenance. ERCOT load forecast missed daily peak by nearly 10% since the temperatures in Dallas area exceeded weather forecast by 5°. Five major unit tripped during peak hours, which caused Emergency Electric Curtailment Plan initiated. First, all available generation, private network and maximum DC Tie supply were deployed. Next, once the system frequency started to decline quickly, LaaRs providing RRS were called at 15:34. Finally, 1000 MW of firm load shedding (rotating outages) were

Table 4.4 Deployment of Laars between 2012 and 2016

Date	Time	Duration (min)	Amount of Response (MW)	Type of Deployment
2012				
7/10/ 2012	20:46	14	198	UF event for frequency near 59.7 Hz that caused some load resources to trip offline
7/30/ 2012	16:03	13	324	UF event for frequency near 59.7 Hz that caused some load resources to trip offline
11/2/ 2012	21:41	11	882	UF event for frequency < 59.7 Hz but of uncertain duration
2013				
1/4/ 2013	9:41	20	572	UF event for frequency near 59.7 Hz that caused some load resources to trip offline
1/8/ 2013	16:40	10	974	UF event for frequency near 59.7 Hz that caused some load resources to trip offline
11/1/ 2013	21:47	10	463	UF event for frequency near 59.7 Hz that caused some load resources to trip offline
2014				
1/6/ 2014	6:53	60	1028	Initially to maintain frequency but then went into EEA level 2 with a system wide deployment
1/18/ 2014	8:41	59	850	UF event for frequency near 59.7 Hz that caused some load resources to trip offline; resulted in EEA level 1 event.
2015				
7/29/ 2015	18:16	10	22	UF event for frequency near 59.7 Hz that caused some load resources to trip offline
2016				
5/1/ 2016	20:20	4	927	UF event for frequency near 59.7 Hz that caused some load resources to trip offline

ordered by ERCOT when the instructions were issued at 16:13. For this event, approximate 2% of load were tripped at distribution level over 2-hour period. When the system frequency ultimately returned back to its normal value, the deployed LaaRs were recalled. Over 1300 MW of LaaRs during the emergency certainly helped ERCOT maintain reliability and reduced the need for rolling blackouts at a larger scale.

Another example of LaaRs deployment is given in Fig. 4.3. When the frequency dropped below 59.7 Hz due to the loss of a large generation unit, those LaaRs providing RRS were immediately released, and then gradually recalled as the frequency was restored back to its normal range.

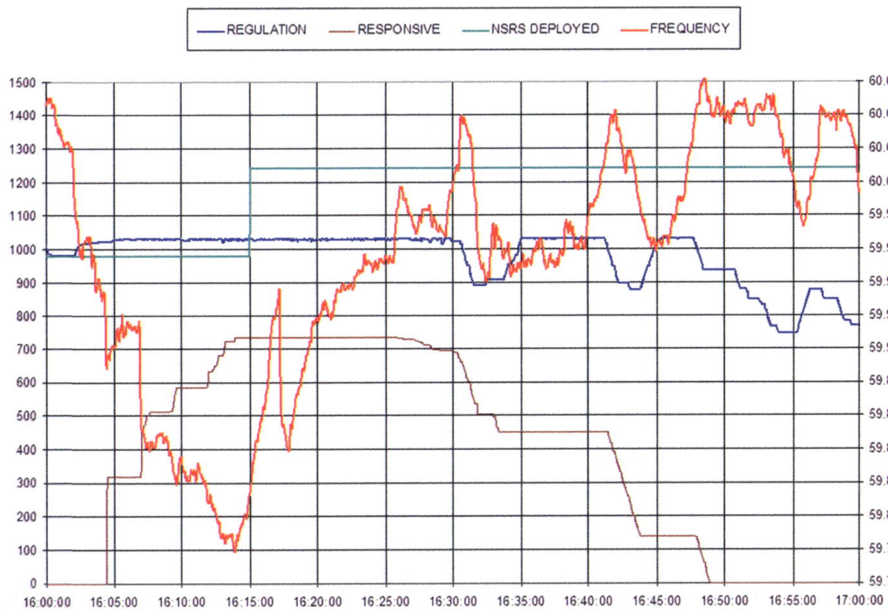

Fig. 4.2 An example of LaaRs deployment

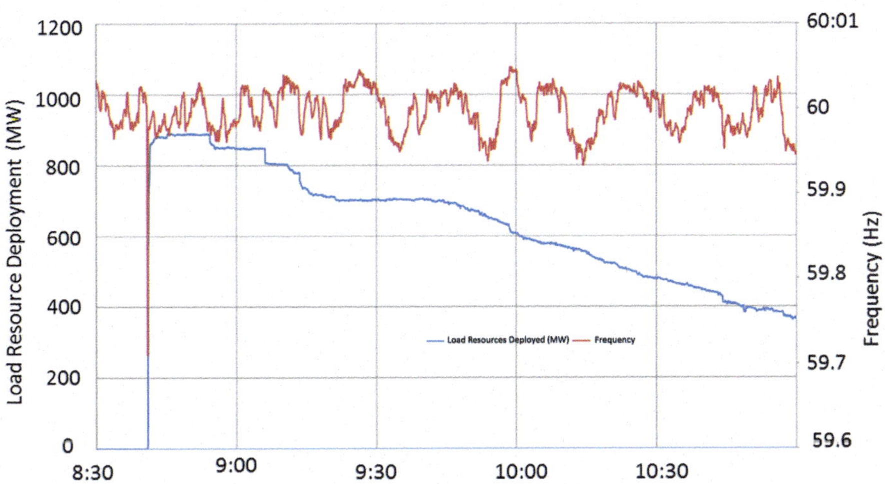

Fig. 4.3 An example of LaaRs autonomous deployment in response to frequency drop below 59.7 Hz

4.6 ERCOT Emergency Response Service (ERS)

4.6.1 ERS History and Requirement

ERCOT has been operating ERS (or preceding Emergency Interruptible Load Service (EILS)) since February 2008. The rational is to develop a service provided by loads (customers) willing to interrupt during an electric grid emergency in return for financial compensation. The Emergency Interruptible Load Service (EILS) was originally developed and implemented in 2007 as a load reduction service with a 10-min ramp requirement to be deployed in the late stages of a grid emergency. Later in 2012 this service was modified to allow certain types of small generators to participate and thus to reflect the broader participation, the service was renamed "Emergency Response Service." Additional significant changes to this service occurred in 2014 when the 30-min ramp service type was implemented as well as the Weather-Sensitive ERS with both 10-min and 30-min ramp requirements. Figure 4.4 shows the historical procured ERS amount and the corresponding cleared price. The trend observed is that the procured ERS gradually increased with a decline in the price. Among the procured ERS amount, a noticeable port was contributed by ERS generators.

ERS is an additional tool for ERCOT Operations, deployed ONLY in the late stages of a grid emergency as last resort prior to firm load shedding (rotating outages). By executing controlled interruption of some prepared customers, a situation of uncontrolled interruption of unprepared end-users at large scale could be prevented. Table 4.5 presents the alert and different levels of emergency defined in

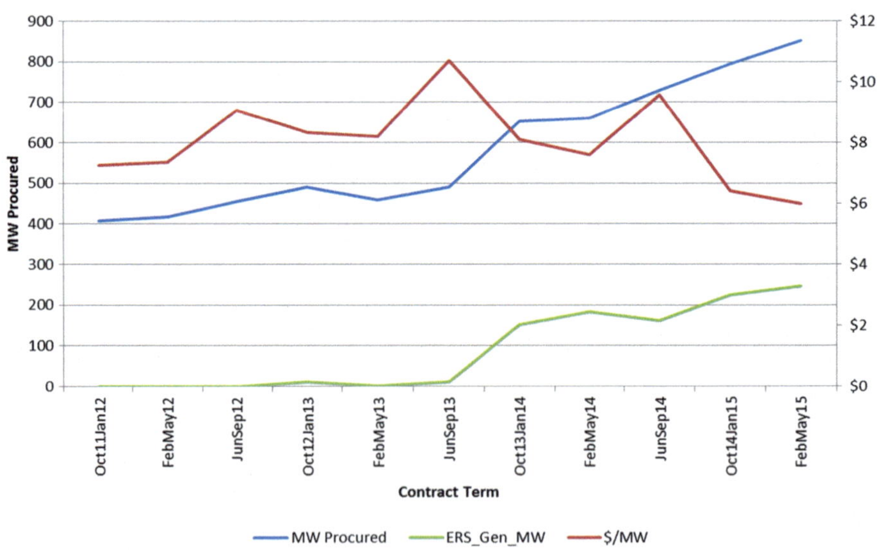

Fig. 4.4 Trend of ERS procurement

Table 4.5 Emergency and actions for response

Event/action	Trigger
ADVISORY: Notice to market participants	Physical responsive below 3000 MW
WATCH: Start reliability must run units, suspend unit testing, deploy replacement and non-spin reserves	Physical responsive below 2500 MW
Energy emergency alert	
Level 1: Dispatch all generation, issue public media appeal, acquire maximum power thru DC ties	Physical responsive below 2300 MW
Level 2A: Deploy LaaRs	Physical responsive below 1750 MW
Level 2B: Deploy ERS resources	Maintain frequency at 60 Hz
Level 3: Instruct transmission owners to shed firm load	Frequency below 59.8 hz

ERCOT operational procedures and the corresponding actions which should be taken to resolve the emergency.

Both individual loads and aggregations of load resources are eligible to participate in ERS. To be qualified for ERS, load resources must meet the following requirements

- Capability of interrupting at least 1 MW of load on 10-min or 30-min notice at any time during the committed hours.
- Have 15-min interval metering or statistically valid sample approved by ERCOT.
- Representation by a QSE with a wide-area network agreement with ERCOT.
- Must have 24/7 operations that can receive the verbal dispatch instruction.
- QSE is responsible for notifying the customer.
- ERS Resources must shed at least 95% of their committed load within 10-min/30-min of QSEs' receipt of the instruction.

One example of deploying 30-min ERS fleet and 10-min ERS is shown in Figs. 4.5 and 4.6, respectively. Each time interval represents 15-minute. In Fig. 4.5, the performance is based on a prorated obligation for the first partial interval only (11.57 min) and therefore prorated obligation for first partial interval was 86.11 MW. The total fleet response was 94.831 MW after 30 min upon the receipt of Verbal Dispatch Instruction (VDI). In the second example, the fleet wide obligation less scheduled unavailability was 495.72 MW and the fleet response was 594.86 MW.

4.6.2 Capacity Demand Curve

To more efficiently procure ERS, ERCOT develops a capacity demand curve for each ERS Time Period, and all ERS products will be procured together within the limits of that curve. According to this demand curve, how much ERCOT will

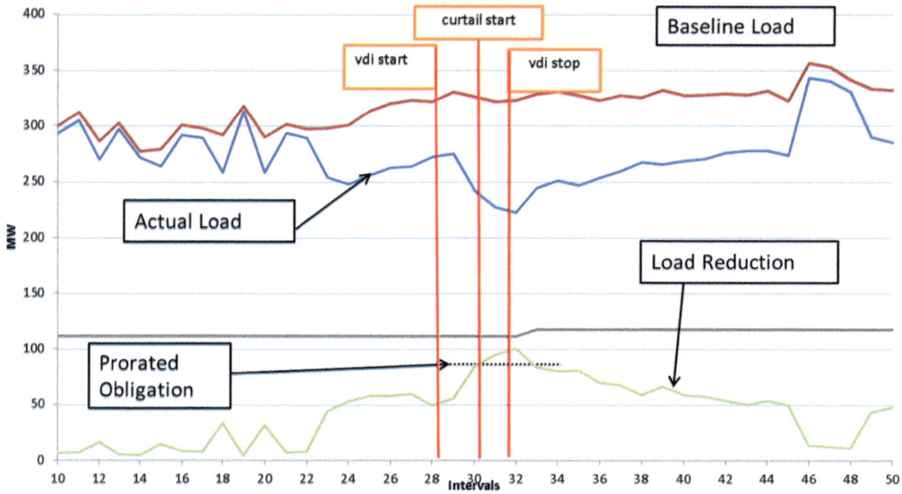

Fig. 4.5 One example of ERS-30 deployment

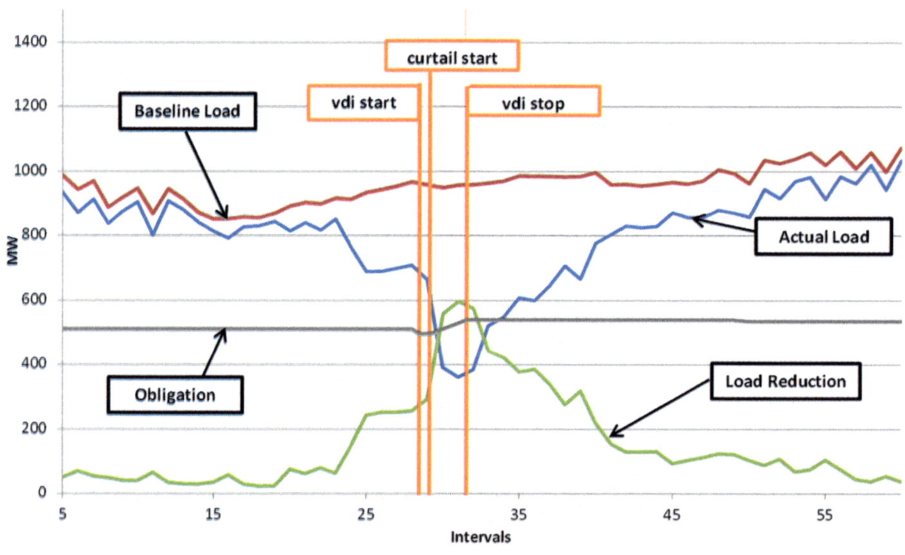

Fig. 4.6 One example of ERS-10 deployment

purchase for ERS depends on the bidding price received. The lower price at which ERS is sold, a higher amount of ERS ERCOT will buy. ERCOT shall maximize the MW procured subject to the expenditure limit for the relevant Time Period.

Each demand curve is derived from the three following parameters, as shown in Fig. 4.7:

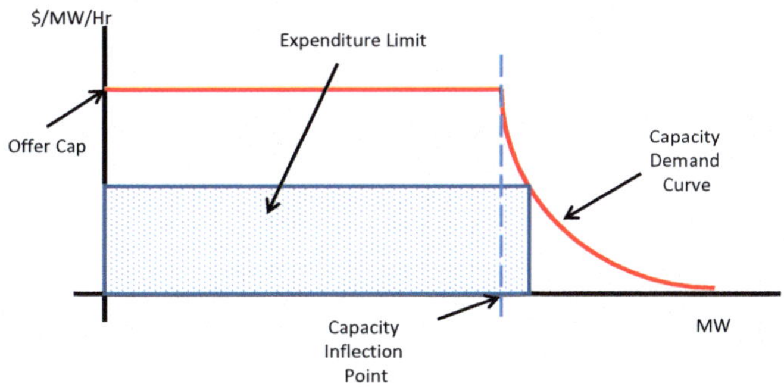

Fig. 4.7 Illustration of capacity demand curve for ERS

1. ERS Offer Cap.
2. ERS Time Period Capacity Inflection Point.
3. ERS Time Period Expenditure Limit.

The ERS offer cap establishes a maximum possible procurement price of \$80/MWh for every ERS Time Period during the ERS budget year. ERCOT will automatically reject any offers above the offer cap. This cap is consistent with the prices historically paid to Loads participating in TDSP Standard Offer Load Management Programs.

P.U.C. Substantive Rule 25.507 restricts ERCOT's ERS expenditures to an annual cost cap of \$50 million. ERCOT will allocate the \$50 million available expenditure within its ERS budget year, which starts with the February through May Standard Contract Term and ends with the October through January Standard Contract Term. No later than 60 days before each new ERS budget year, ERCOT will make an initial allocation of the annual expenditure limit to each ERS Time Period in each ERS Standard Contract Term based on the expected risk of reaching an Energy Emergency Alerts (EEA) in that ERS Time Period, in accordance with the formula detailed below. ERCOT will assign a high (H), moderate (M), or low (L) risk designation to each ERS Time Period and will assign a risk-weighting factor (a value from 1 to 10 with 10 being the highest risk value) for each risk designation. ERCOT's risk assessment will consider a number of factors, including, but not limited to, forecasted operating reserves, forecasted Load, and Resource outage information. 60 days before each ERS Standard Contract Term begins, ERCOT will update the allocation for each remaining ERS Time Period in the budget year to reflect updated forecasts and any expected remaining funds from ERS Standard Contract Terms within the same ERS budget year. Any funds remaining at the end of an ERS budget year will not be carried forward into a new ERS budget year.

For each ERS Time Period, the expenditure limit is calculated as follows:

$$\textit{Expenditure Limit}_{TP} = \textit{Annual Expenditure Limit} \times \begin{matrix}\textit{Expenditure} \\ \textit{Limit} \\ \textit{Allocation} \\ \textit{Factor}\end{matrix}_{TP} \tag{4.1}$$

where

$$\begin{matrix}\textit{Expenditure} \\ \textit{Limit} \\ \textit{Allocation} \\ \textit{Factor}\end{matrix}_{TP} = \left[\begin{matrix}\textit{Risk} \\ \textit{Weighting} \\ \textit{Factor}\end{matrix}_{TP} \times \#\mathrm{h}_{TP} \times \begin{matrix}\textit{Offer} \\ \textit{Cap}\end{matrix}\right] \div \left[\sum_{TP} \begin{matrix}\textit{Risk} \\ \textit{Weighting} \\ \textit{Factor}\end{matrix}_{TP} \times \#\mathrm{h}_{TP} \times \begin{matrix}\textit{Offer} \\ \textit{Cap}\end{matrix}\right]$$

The capacity inflection point establishes the point on the capacity demand curve where capacity can only be procured at an offer price less than the ERS Time Period offer cap while respecting the expenditure limit for that ERS Time Period. The capacity inflection point for each time period is calculated as follows:

$$\textit{CapInflection Point}_{TP} = \textit{Expenditure Limit}_{TP} \div [\#\mathrm{h}_{TP} \times \textit{Offer Cap}] \tag{4.2}$$

The standing ERS Time Periods are as follows (all times are Central Prevailing Time) (Table 4.6).

The highest offer accepted for an ERS Time Period will set the clearing price for all ERS Resources cleared in that ERS Time Period. All ERS service types will be procured using a common ERS capacity demand curve for each ERS Time Period and the highest offer accepted for an ERS Time Period will set the clearing price for all ERS service types.

Table 4.7 below provides hypothetical calculations of the expenditure limits and capacity inflection point for each ERS Time Period in each budget year.

Table 4.6 Standing ERS time periods

Time period name	Time period hours
Business hours 1 (BH1)	Hours Ending 0900–1300 (8:00:00 a.m. to 1:00:00 p.m.)
	Monday through Friday except ERCOT holidays.
Business hours 2 (BH2)	Hours Ending 1400–1600 (1:00:00 p.m. to 4:00:00 p.m.)
	Monday through Friday except ERCOT holidays.
Business hours 3 (BH3)	Hours Ending 1700–2000 (4:00:00 p.m. to 8:00:00 p.m.)
	Monday through Friday except ERCOT holidays.
Non-business hours (NBH)	All other hours

Table 4.7 ERS time period expenditure limit allocation and capacity inflection point calculations

		Risk level	Risk weighting factor (a)	Time period hours (b)	Price Cap (c)		(a)*(b)* (c)	Expenditure limit allocation factor (%)	ERS time period expenditure limit		Capacity inflection point (MW)
Feb–May	bh1	L	1	430	$	80	34,400	3.05	$	1,527,423	44.4
	bh2	L	1	258	$	80	20,540	1.83	$	916,454	44.4
	bh3	L	1	344	$	80	27,520	2.44	$	1,221,938	44.4
	Nbh	L	1	1871	$	80	149,580	13.29	$	6,646,064	44.4
					$				$		
Jun–Sep	bh1	L	1	420	$	80	33,600	2.98	$	1,491,901	44.4
	bh2	H	10	252	$	80	201,600	17.90	$	8,951,407	444.0
	bh3	H	10	335	$	80	268,800	23.87	$	11,935,209	444.0
	Nbh	L	1	1920	$	80	153,600	13.64	$	6,820,119	44.4
					$				$		
Oct–Jan	bh1	L	1	420	$	80	33,600	2.98	$	1,491,901	44.4
	bh2	L	1	252	$	80	20,160	1.79	$	895,141	44.4
	bh3	L	1	335	$	80	26,880	2.39	$	1,193,521	44.4
	Nbh	L	1	1945	$	80	155,600	13.82	$	6,908,923	44.4

4.6.3 Baselines for ERS Loads

One difficulty in estimating the performance of ERS load when being deployed is that a robust and accurate baseline method is required. The baseline is a way of statistically estimating an ERS Resource's electricity usage under business-as-usual conditions in the absence of a deployment. The baseline is designed to yield interval-by-interval load estimates for the ERS Load to provide the most accurate possible benchmark for evaluating the performance of the ERS Load in a deployment event. This estimate can then be compared to the ERS Load's interval meter data from an ERS deployment event to determine its performance throughout the ramp period and Sustained Response Period.

ERCOT will assign each individual or aggregated ERS Load to its own unique baseline type. In some cases, an ERS Load may be qualified for multiple baseline types and may choose its preferred assignment from the options offered by ERCOT. The baseline will be used to verify the ERS Load's performance as compared to its contracted capacity during an ERS deployment event and thus is a key determinant in ERCOT's measurement and verification of the ERS Load's availability.

Figure 4.8 illustrates one example of using the baseline to evaluate the performance of ERS deployment, where the solid green line represents the baseline, the dot black line is the metered load, and the shadow area is expected load shape after 1 MW contracted load reduction is considered. The performance in an ERS deployment event is measured by comparing actual load data to the baseline. Interval performance factors are calculated by comparing actual load to the baseline for each 15-minute interval in the event and Event Performance Factor is the average of

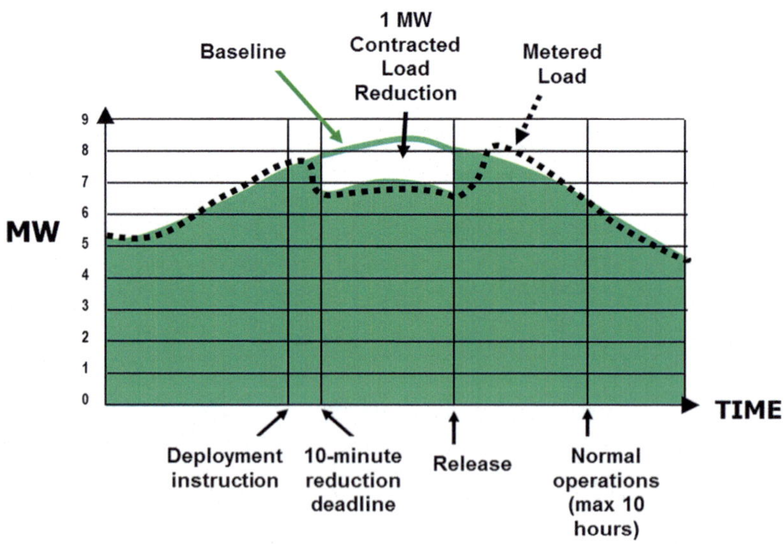

Fig. 4.8 Baseline and actual load for an ERS event

all interval performance factors. If the event performance factor is less than 0.95, ERS Resource is subject to 6-month suspension.

A baseline assignment applies to the ERS Load for all Time Periods and Contract Periods within a Standard Contract Term. ERS baseline methodologies fall into two categories: *default* or *alternate*. ERCOT will evaluate interval-metered Load from prospective ERS Loads to determine their baseline assignments. All sites within an aggregated ERS Load must be assigned to the same baseline type (one of the three default baseline types or the alternate baseline).

4.6.3.1 Default Baseline

If requested by QSE, ERCOT, using historical meter data, will attempt to establish a baseline model for an ERS Resource under one of three methodologies:

1. The Regression Model applies the methodology used for the development and support of ERCOT's Load Profiles to develop an unadjusted ERS Load-specific model. In addition to the ERS Load's interval-meter data (including historic data plus data for the time periods preceding and following the deployment event), the model inputs for this default baseline model fall into three general categories: weather-related variables (e.g., temperature, dew point, wind speed, and cloud cover); calendar-related variables (e.g., day of week, holiday, and season); and daylight/darkness variables (sunrise and sunset time). ERCOT, at its sole discretion, may use other data variables in the baseline formula if ERCOT determines the additional data will enhance the accuracy of the baseline.
2. The "Middle 8-of-10 Preceding Like Days" Model uses the ERS Load's actual interval-meter data for the 10 "like" days (weekdays or weekends/holidays) immediately preceding the day of an ERS deployment event, discards the highest Load day and the lowest Load day of the 10, and creates an unadjusted baseline by averaging the Load of the remaining 8 days.
3. The Matching Day Pair Model is based on identifying the sets of 2 days over the preceding 12 months in which the ERS Load's actual interval-metered usage most closely matches its actual Load data for day before and the day of the ERS deployment event. This methodology selects the 10 closest-matched day pairs and averages them together by interval to create an unadjusted baseline.

The primary goal of a default baseline is to accurately estimate an ERS Load's level of electric energy usage under "business as usual" conditions. Depending on the default baseline type, analysis of at least 12 months of historic interval meter data may be necessary for ERCOT to determine whether an ERS Load can be modeled accurately under a default baseline.

ERCOT will apply an event-day adjustment to the unadjusted baseline Load estimates for any of the three default baseline types described above. An event-day adjustment is a scalar adjustment based on the ERS Load's actual Load for the hours preceding the declaration of an EEA. An event-day adjustment may change the baseline up or down. QSEs and ERS Loads may use these methodologies for estimating their own specific baseline for a particular day or event. For aggregated ERS Loads, ERCOT may develop either a single baseline model at the aggregate level or multiple baseline models for individual Sites or subsets of Sites within the aggregation. If ERCOT develops the model at the Site and/or subset level, ERCOT shall establish the default baseline for the aggregation by summing the baselines of the individual Sites and/or subsets of Sites in the aggregation. ERCOT shall verify the performance at the ERS Load level.

4.6.3.2 Alternate Baseline

Some loads do not have enough predictability to allow ERCOT to create a default baseline model; for example:

- Fluctuating or batch-process loads.
- Loads with unpredictable downtimes or irregular schedules.

ERCOT will assign such loads to the Alternate Baseline performance evaluation methodology. It is not a true "baseline" because it does not attempt to estimate the resource's load in the absence of the deployment. In an ERS deployment event, load must stay at or below its declared Minimum Base Load throughout the event, i.e., the MW commitment of ERS should be less than average hourly load minus declared Minimum Base Load. In Fig. 4.9, if ERS Resource offer is awarded at 20 MW, when deployed, the ERS Resource's load level must remain at or below its declared Minimum Base Load of 4 MW throughout the event.

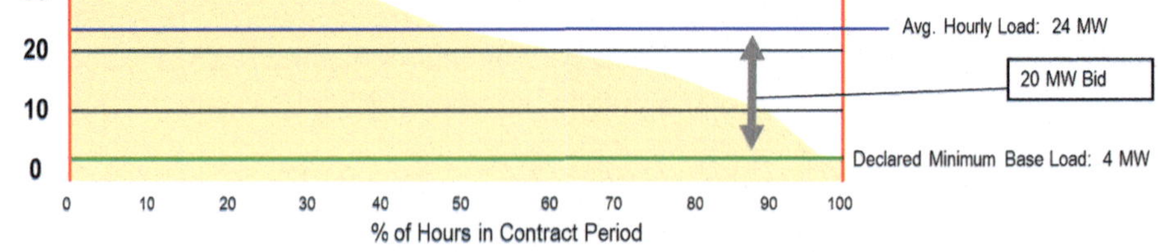

Fig. 4.9 Alternate baseline of ERS

4.7 Load in SCED

Consumers have had the ability and incentive to change consumption in response to real-time energy prices. Demand responding to price (as supply currently does) is important as it is a crucial component of an efficient market design.

Under the current Texas Nodal market construct, which was launched in December 2010, consumers wishing to respond to energy prices have found the market significantly more difficult than in the previous Zonal market. While Load Zone Locational Marginal Prices are now published with each run of SCED, an inherent characteristic of the Nodal market is that the actual prices paid by loads are not calculated and known until after the end of a 15-min Settlement interval. The mechanics of this change results in prices that cannot be relied upon as an indicator for final settlement nor are these prices provided with enough advance notice to allow for effective response. This is because there is a minimum of three and a practical limit of approximately 3 SCED runs per 15-min settlement interval. The price results of each SCED run associated with a settlement interval are averaged to determine the final settlement price. This leads to a reduction of certainty in economic load response as the initial published price that elicits a change in consumption may have minimal impact on the final settlement price. Further, the first published price will be known only 5 min in advance of the start of the settlement interval, reducing the time available to respond.

A proposed solution is outlined here to overcome these barriers to price response in the current market design by incorporating demand response into the SCED dispatch engine. Figure 4.10 illustrates the resulting real-time price when price elasticity of the end-use loads is taken into consideration. The bids can be submitted by two primary types, which include suppliers (bid in amount they can provide and what they want to charge for it) and demanders (bid in amount they wish to use and how much they are willing to pay for it). After the bidding process is stopped and supply and demand bids are arranged, the meeting point of the two curves is the "clearing price and quantity" of the auction. In this case, demand becomes elastic and it can be allowed to set the real-time price.

Load participation in SCED could contribute toward the goals of increasing the amount of competitive demand response in the ERCOT energy markets, broadening the participation in the energy markets by customers of all sizes, and increasing the efficiency of the ERCOT energy and ancillary markets.

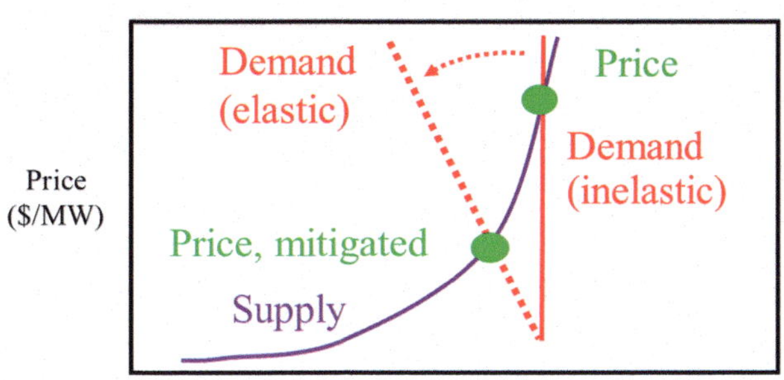

Fig. 4.10 Real-time price: price elasticity of the end use loads

4.7.1 Benefits of Load Participation in the Real-Time Market

Load participation in SCED broadens the electric market by enabling consumers to participate as economically-dispatched demand response (DR). This has the potential to increase market efficiency through price elasticity of demand, lower overall costs to consumers, and expand the pool of assets available to ERCOT to ensure grid reliability.

Load participants will each determine whether and how best to offer their DR capability into the market. Some are likely to submit offers that represent their opportunity cost or value of lost load – that is, energy price offers that may be significantly higher than those reflecting generators' marginal costs. This approach is capable of improving ERCOT's energy-only market design by providing shortage prices when appropriate, based on market offers that reflect the shortage conditions.

Some load participants can be expected to offer into SCED as a result of being procured to provide an AS in the ERCOT Day-Ahead Market (DAM). SCED dispatch of Load Resources providing AS will be truly economic—reducing the need to rely solely on the alternative deployment methodologies inherent in DR dispatch today, which can perversely suppress prices during shortage conditions.

Furthermore, load participation in SCED can help to remedy a characteristic of the ERCOT Nodal market that has been identified by loads as an impediment to economic demand response. In the Zonal market, loads had the ability to respond to price signals because of ex ante price transparency to the end user, as Zonal energy clearing prices were posted at least 5 min prior to the start of the operating interval. This could also expand the opportunity for load participation in SCED by allowing aggregations of distribution-level loads to become qualified as a new type of Load Resource. Expanding the ERCOT energy market to these customers would further increase competition, and help to tap the DR potential of consumers newly empowered from the mass deployment of advanced meters.

In short, enabling load participation in SCED creates a platform that significantly improves the economic convergence of supply and demand in the ERCOT electricity market.

4.7.2 Eligibility of Loads in SCED

Types of loads eligible to participate in SCED include

(a) Transmission Connected Loads – Transmission connected loads, typically large industrial consumers, can readily be incorporated into the ERCOT Network Operations Model at the transmission bus where they receive service. These loads can potentially be used for both power balancing and congestion management, the simultaneous primary duties of the SCED engine.
(b) Distribution-level single loads – Distribution-connected individual loads can be assigned to specific Resource Nodes on the Network Model via consultation

between the TDSP and ERCOT. Once assigned to a Resource Node they will have shift factors calculated that will enable them to be dispatched by SCED for congestion management.

(c) Aggregations of Distribution-level loads – Participation by aggregated distribution-level Loads which is entitled "Aggregated Load Resources."

4.7.3 Modifications to SCED

Load Resource participation in SCED will incorporate many of the basic concepts and functionality already included in the Nodal market design. Loads would be eligible, through their QSEs, to submit energy-only offers directly into SCED or to submit those offers in conjunction with an obligation to provide Ancillary Services.

4.7.3.1 Mathematical Formulation of SCED

The detailed mathematical formulation of SCED with load resources modeled is given as follows.

Objective: Determine the Base Points for Generation Resource (MW to supply) and Base Points for Load Resources with Demand Bid (MW to consume) such that the following function is minimized:

$$\min_{G,LR}\left\{\sum_{G} C_{G}^{Offer}\left(BP_{G}^{Supply}\right)-\sum_{LR} C_{LR}^{Bid}\left(BP_{LR}^{Demand}\right)\right\} \tag{4.3}$$

This is a quadratic objective function as current SCED – the only change is inclusion of additional quadratic terms representing bid to buy from LR.

Subject to following constraints: *Power Balance*:

$$\sum_{G} BP_{G}^{Supply} = GTBD + \sum_{LR} BP_{LR}^{Demand} - \sum_{LR} NPF_{LR}^{Demand} \tag{4.4}$$

$$GTBD = \sum_{G} TelemMW_{G} + Other\_Terms \tag{4.5}$$

BP_{G}^{Supply} is the SCED Base Point instruction to Generation Resource (G) on how much MW to *supply*. It is expected that the Generation Resource Telemetered MW output ($TelemMW_{G}$) will be equal to its instructed Base Point (absent any Regulation deployment or governor response).

BP_{LR}^{Demand} is the SCED Base Point instruction to Load Resource (LR) on how much MW to *consume*. It is expected that the Load Resource Telemetered MW actual consumption (or Net Power Flow (NPF_{LR})) will be equal to its instructed Base Point.

OtherTerms include expected 5-min Load Ramp, Operator Offset, Regulation Bias, etc.

Resource Constraints for Generation Resources:

$$LRL_G \leq LSL_G \leq LASL_G \leq LDL_G \leq BP_G^{Supply} \leq HDL_G \leq HASL_G \leq HSL_G \leq HRL_G \tag{4.6}$$

LDL_G and HDL_G are each Generation Resource's (G) dispatch limits. These dispatch limits are calculated using the submitted telemetered output MW ($TelemMW_G$) as reference and also uses the submitted telemetered values of ramp rates ($RR^{Up}{}_G$, $RR^{Down}{}_G$), LSL_G and HSL_G and the ERCOT calculated $LASL_G$ and $HASL_G$.

Resource Constraints for Load Resources:

$$LRL_{LR} \leq LPC_{LR} \leq LASL_{LR} \leq LDL_{LR} \leq BP_{LR}^{Demand} \leq HDL_{LR} \leq HASL_{LR} \leq MPC_{LR} \leq HRL_{LR} \tag{4.7}$$

LDL_{LR} and HDL_{LR} are each Load Resource's (LR) dispatch limits. These values are calculated using the submitted telemetered actual consumption MW (Net Power Flow (NPF_{LR})) as reference. These dispatch limits are calculated using the submitted telemetered actual consumption MW (Net Power Flow (NPF_{LR})) as reference and also uses the submitted telemetered values of ramp rates ($RR^{Up}{}_{LR}$, $RR^{Down}{}_{LR}$), LPC_{LR} and MPC_{LR} and the ERCOT calculated $LASL_{LR}$ and $HASL_{LR}$.

Transmission Constraint:

$$\sum_G SF_G^c \times BP_G^{Supply} + \sum_{LR} SF^c_{LoadZone\_LR} \times \left(NPF_{LR} - BP_{LR}^{Demand}\right) \leq Rating^c - Flow_{SE}^c + \sum_G SF_G^c \times SE\_MW_G^{Supply} \tag{4.8}$$

4.7.3.2 Solving SCED Problem

The current SCED process uses a two-step methodology that solves for transmission congestion by applying mitigation to resolve Non-Competitive Constraints. This process evaluates offers and output schedules only from On-Line Generation Resources and dispatches to the total generation requirement determined by Load Frequency Control (LFC), subject to transmission constraints. The first step determines the Reference Locational Marginal Price (LMP) while observing only the limits of Competitive Constraints, the list of which is determined by the Constraints Competitiveness Tests (CCT). The second step then produces the base points, shadow prices, and LMPs using EOCs, which are either provided by QSEs for all on-line Generation Resources or created by ERCOT, while observing the limits of both Competitive and Non-Competitive Constraints. In this step the EOCs are now capped by either the Reference LMP as determined in the first step or the appropriate Mitigated Offer Cap and bounded at the lesser of the Reference LMP or the appropriate Mitigated Offer Floor.

To accommodate Load Resource participation, SCED step one would be modified to include evaluation of Load Resource EOCs submitted by QSEs or created by ERCOT. ERCOT will create EOCs only for those LRs carrying an AS Resource Responsibility for that time period.

Step two would then use EOCs from both Generation and Load Resources subject to security constraints (including both Competitive and Non-Competitive Constraints) and other Resource constraints. This step will be used to produce Base Points, Shadow prices, and LMPs for both Generation and Load Resources while applying the prevailing Offer Cap and offer floor mitigation rules.

The current SCED process uses resource parameters, such as ramp rate, low sustained limit (LSL), high sustainable limit (HSL), low dispatch limit (LDL), high dispatch limit (HDL), and other attributes. Load Resources should be able to use all of these parameters to customize their participation in SCED. For example, if the MW consumed by a process changes with time, either rapidly (such as a steel mill) or slowly (such as a shift-based production process), the QSE representing the Load Resource could telemeter to ERCOT a changing HSL, so that SCED only uses the energy available for dispatch at that time.

SCED will also require modification to allow energy deployed by a Load Resource to be fed back into SCED's ongoing algorithm calculating generation to be dispatched. This will prevent SCED from potentially causing volatile oscillating deployments of Load Resources with each SCED execution.

Adoption of a minimum run-time parameter could assure Load Resources participating in SCED of a payment for their energy deployment that more accurately approximates their opportunity cost or value of lost load. In the absence of this or an alternative energy payment mechanism, LRs would be guaranteed to receive the LMP for only a single SCED execution. This lack of certainty for the energy payment could be a deterrent to Load participation.

4.7.4 Compensation of Demand Response in the Energy Market

Various compensation methods have been proposed to discuss how to pay ERS for their deployment. FERC's Order 745[4] requires the FERC-jurisdictional Independent System Operators (ISOs) and Regional Transmission Organizations (RTOs) to compensate demand response at the full Locational Marginal Price (LMP). A number of stakeholders in the federal docket, including the members of the

[4]http://www.ferc.gov/industries/electric/indus-act/demand-response.asp

ISO-RTO Council, had submitted comments encouraging a different approach that would compensate demand response at LMP minus “G,” where G represents a value equivalent to the generation costs avoided by the customer as a result of the demand response deployment. The following compensation examples are presented to illustrate the difference between two options.

4.7.4.1 Option 1: Settlement Concept: LMP minus LMPz

In this scenario, Load Resources are paid for deployed energy at the Real-Time Settlement Price Point (RTSPP). This means that whenever SCED determines that a particular LR is required to solve, that LR receives a base point above zero and is therefore eligible to be paid the amount of energy it deployed times the RTSPP.

To ensure loads would not be paid to provide energy which they did not have available to sell, the LR’s response must be added to its Load Serving Entity’s (LSE’s) load zone energy consumption. For example, assume that SCED dispatches all of an 82 MW offer at $2700 from a LR that is currently consuming 100 MW. As a result, the LMP at the resource node was $2700 and the zonally weighted average price (LMPZ) was $200. Then assume that the LR deployed the offered 82 MW, leaving 18 MW online. The settlement results are as follows:

1. The QSE representing the LR’s demand response would be paid 82 times $2700, or $221,400.
2. The LSE’s settled load for this resource would be 18 MW (its actual consumption during the deployment) + 82 MW (its demand response added back into its consumption); therefore, the LR’s QSE would be charged 100 MW multiplied by the $200 LMPZ, or $20,000. The QSE could have other positions at other settlement points, but these are ignored for simplicity.
3. The QSE’s simplified settlement statement for this day would total $201,400.

4.7.4.2 Option 2: Alternative Approach: Full LMP

While many parties believe that loads participating in the energy market should be paid the LMP minus the retail rate (or the generation/supply component of the retail rate), the alternative payment structure is to pay loads the LMP. In order to balance supply and demand, generators can increase supply or loads can decrease demand. At times when generators perform balancing, they are paid LMP so it is equitable that when loads perform balancing they should be paid LMP. Recently, FERC ordered that ISOs, under its jurisdiction, establish processes to pay loads the LMP when it is cost effective to do so. Such an approach ensures that ratepayer costs to manage the grid are least cost. Additionally, LMP creates a level-playing field in an open market rather than one constrained by an administrative restraint on the market.

Paying LMP is also less complex as determining a retail rate or generation component for each load that participates can be very complicated. Paying LMP would invite participation from loads in all zones rather than only in zones where LMP is high and the retail rate is low.

When loads are paid LMP minus a retail rate, REPs have no risk and are held harmless. That is good for REPs; however, that reduces innovation in the market. If REPs had some risk, they would work with their retail portfolio to lessen that risk and innovation would result. In addition, during the periods when LMPs are high and load participation most likely, REPs would actually make money under a LMP-G approach because they could sell the unused commodity at the generally higher LMP price.

4.7.5 Performance of SCED Dispatch for Load

Figure 4.11 shows one example of energy dispatch from SCED for Controllable Load Resources (CLRs). To be eligible for SCED dispatch, the load resources are either existing or new single-site CLRs or Aggregate Load Resources (ALRs) composed of multiple sites within single ERCOT Load Zone. LRs in SCED will submit bids to buy (not offers to sell) and these bids will reflect LR's willingness to consume up to a specified 5-min Load Zone LMP. Accordingly, the bid will modify the SCED demand curve and have ability to set price. It can be seen in Fig. 4.11 that the dispatch base-points were created by adding SCED dispatch quantity to the

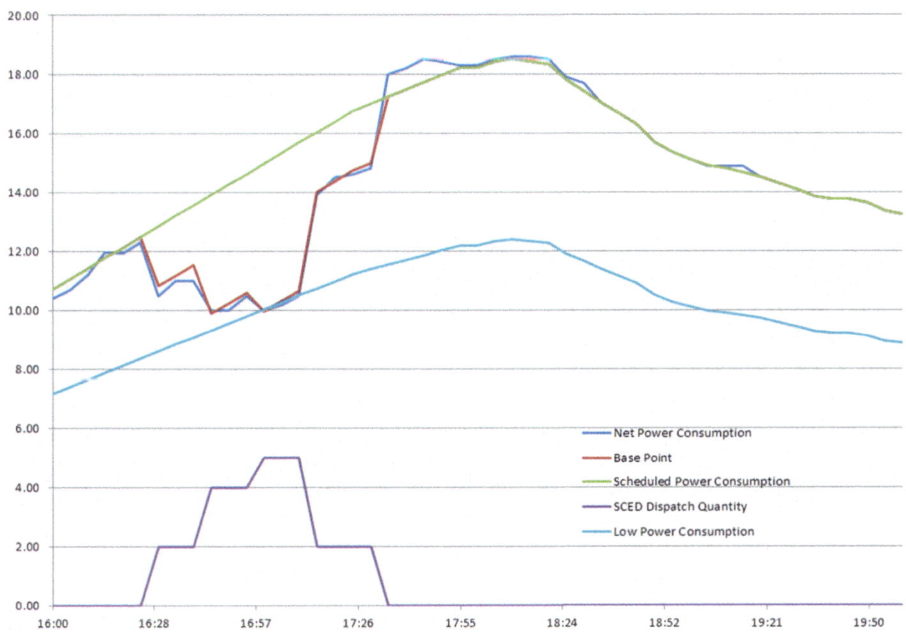

Fig. 4.11 Energy dispatch from SCED for CLRs

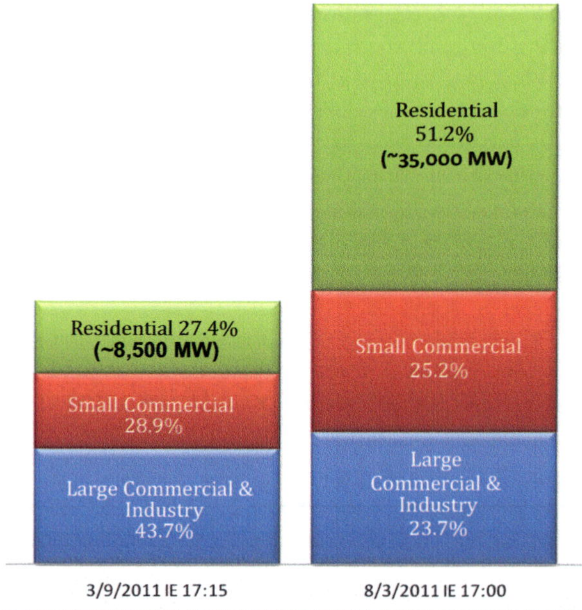

Fig. 4.12 Load composition

scheduled power consumption. For this particular case, Load Resources were capable of following 5-minute SCED base point instructions and also complied with the limits for low energy consumption.

4.8 Summary

Demand response programs in ERCOT provide opportunities for demand-side resources to defray their electricity costs and provide reliability benefits to the ERCOT electric grid.

Electric power system is designed to support highest peak during the peak day of the year. The peak load is due to coincidental usage and is driven by the end-users. Figure 4.12 shows the impact of weather on load by customer type at ERCOT. On the left side, the ERCOT load at 5:15 PM on March 9, 2011 (Wednesday) was 31,262 MW and the temperature in Dallas was 64 ° F. For such a load level, a large number of generators were shut down as idle. On the right side is a peak-load of 68,416 MW at 5:00 PM on August 3, 2011 (Wednesday) and the temperature in Dallas was 109 ° F. To meet this high load demand, both base-load and peak-load generation need to be online. In an unstructured utility, capital investments are driven up by the need to serve this peak load so that electrical infrastructure is underutilized for the majority of the time. In a power market environment, the decision of building new generation is a market behavior but is still largely

Table 4.8 Summary of demand response at ERCOT at peak-load hours

Service	MWs	Notes
Load resources providing responsive reserves	≤1400	• "LaaRs" -- procured on an hourly basis in the Day-Ahead Market for next Operating Day • Deployed via 10-min instruction or instantaneously based on grid frequency • Determined in 4 h blocks, LRs can still provide up to 60% of the RRS requirement
Emergency response service (ERS)	~635	• DR and DG assets with 10-min or 30-min ramp periods • Also open to weather-sensitive loads such as residential AC aggregations • Procured three times per year (4-month strips), with 4 Time Periods within each contract term
TDSP load management programs	~240	• Commercial & Industrial Loads procured for summer peak hours only • Dispatched concurrently with ERS through agreements between ERCOT and TDSPs

influenced by the prospect of future load shape. An efficient demand response is a prerequisite to the efficient market operation, especially during peak-load hours.

Demand responses are expected to come with the benefits relating to peak load shaving or shifting. Table 4.8 provides an estimated amount of demand responses which could be available at peak-load hours at ERCOT. The total amount is 2200+ MW and this represents approximately 3.3% of ERCOT summer peak. This capacity could be invaluable to reduce the cascading failure risk when the grid is stressed by exceptionally high load demand. In addition to the avoided investment in generation, other benefits include short run marginal cost savings, the value of CO_2 reductions and avoided investment in the distribution and transmission and distribution network.

To regulate the frequency on an interconnected power system, it is necessary to maintain a balance between generation and load plus losses. Traditionally the load has not actively participated in the balancing process. If a user turns on the light switch when the power system is in a stressed condition, the power will be still served. Since the load does not actively participate, it is necessary for the generators to account for all imbalances. To keep a balance between load and generation every instant is currently achieved by adjusting the output of large central generation plants. Greater reliance on generation resources exposes customers and investors to more volatile prices, and will put pressure on grid transfer capabilities and make it harder for system operators to align generation and customer demand levels in real time at a high penetration of renewable generation. For these reasons, it is increasingly apparent to explore more options on the demand side of the power system in the balancing process, and to enlist customers and their agents as ancillary service providers.

ERCOT has made significant progress in developing policy and market rules to accommodate demand responses in market. Participating in ancillary services on the daily base helps load resources to reduce electricity costs and provide additional revenues while posing a relatively low risk to their operations. Nowadays, load resources can contribute to up to 60% of RRS requirement at ERCOT, which is valued at tens of million US dollars annually. For load resources, additional cost may incur relating to the investment on telemetry and reschedule of their operations so as to meet the qualification requirement in the provision of AS. To participate in these programs, load resources have to weigh the economic benefit of participation against the cost. The economic benefits of participation should provide a sufficient incentive to load resources in order to develop a successful program. The experiences at ERCOT show that AS market is one of most attractive services to load resources as the capacity-based payment received from the service can offer a steady revenue stream to offset the cost of participation. Moreover, as these resources were rarely deployed in the past for severe contingencies, the impact of potential deployment on customers is acceptable as long as there is the flexibility available to interrupt the load temporally. The success of load participation in AS is also being seen in other regions.

ERS offers load resources another great opportunity for participation as its response time and qualification requirement are less restrictive compared to AS products. For the Weather-Sensitive ERS, the minute ramp requirements are 10 or 30 min. ERS is reserved as a last resort to prevent large-scale blackout of the grid prior to involuntary load shedding. The compensation for load resources in provision of ERS is capacity based, regardless of how many times they are deployed and how much the required reduction is.

Load participation in SCED promises to level the playing field for supply and demand in the ERCOT energy market. It would empower energy consumers to receive fair compensation for providing demand response and would allow the energy-only ERCOT market to predictably clear at prices that equate to the value of lost load – and thus truly reflect scarcity or shortage conditions. Attracting robust levels of Load Resource participation in SCED may require regulatory action affirming a load's right to submit offers and be compensated based on its LMP. While the benefits of load participation in SCED are apparent, the participation of demand response in SCED at ERCOT is still limited at this moment due to lack of a stable source of revenue. In the long term, the cost and benefit of allowing SCED dispatch of load should be equally considered and weighed so that demand response can play a more significant role in setting the real-time energy price.

Demand-response resources can provide numerous benefits to power systems, particularly for a future grid where a large amount of renewable generation will be integrated. However, how to integrate demand response in a market environment is a challenging new area for most utilities and end-users. While significant progresses have been made towards these challenging goals, these challenges are also evolving and growing. It is very useful to build on lessons from other power markets and system operators that have some years of experience in this arena. The experiences at ERCOT clearly show that DR can be a significant and valued resource to the grid if the response characteristics and operation constraints of load resources can be accommodated. Equally important are the region-specific market rules and policies to open the generation-dominant market to these load resources.

Further Reading

1. Kundur, P. (1994). *Power system stability and control*. New York: McGraw-Hill.
2. Biggar, D. R., & Hesamzadeh, M. R. (2014). *The economics of electricity markets*. Hoboken: Wiley.
3. T. B. Group, S. & Freeman, & Global Energy Partners, L. (2009). *A National Assessment of demand response potential*. Federal Energy Regulatory Commission.
4. *Load participation in the ERCOT nodal market*. ERCOT, 2015.
5. *ERCOT protocols—Nodal*, ERCOT, 2018. Retrieved from http://www.ercot.com/mktrules/nprotocols
6. ERCOT. (2014). *EMERGENCY RESPONSE SERVICE procurement methodology*. ERCOT.
7. ERCOT. (2017). *ERCOT Methodologies for determining minimum ancillary service requirements*. ERCOT.
8. *Demand response strategy*. PJM interconnection, 2017.
9. *Demand response in the Midwest ISO an evaluation of wholesale market design*. The Brattle Group. 2010.
10. *International review of demand response mechanisms*. The brattle Group, 2015.
11. *Demand response as a power system resource program designs, performance, and lessons learned in the United States*. The Regulatory Assistance Project (RAP), May 2013.
12. Du, P., & Lu, N. (2011). Appliance commitment for household load scheduling. *IEEE Transactions on Smart Grid, 2*(2), 411–419.
13. Liu, C., & Du, P. (2018). Participation of load resources in day-ahead market to provide primary-frequency response reserve. *IEEE Transactions on Power Systems, 33*(5), 5041–5051.
14. Li, W., Du, P., & Lu, N. (2017). Design of a new primary frequency control market for hosting frequency response reserve offers from both generators and loads. *IEEE Transactions on Smart Grid, 9*(5), 4883–4892.
15. Du, P., Baldick, R., & Tuohy, A. (2017). *Integration of large-scale renewable energy into bulk power systems*. Berlin: Springer International Publishing.

Chapter 5
Integrated Demand Response in the Multi-Energy System

5.1 Introduction

Faced with the serious environmental challenges and energy shortage brought on by the massive consumption of fossil fuels, it has become a global consensus to develop renewable and sustainable energy [1, 2]. To solve this worldwide problem, smart grid is actively constructed to facilitate the high penetration of renewable generation [3, 4]. However, due to the limited accommodation capability of smart grids, renewable energy curtailment is still a severe issue in existing energy systems. Therefore, the concept of energy internet is emerging at the historic moment [5]. In the book "The Third Industrial Revolution," Jeremy Rifkin presents that by integrating various forms of energy, the smart grid-centered energy internet will change the conventional energy consumption patterns and stimulate the development of renewable and sustainable energy [6]. In 2015, the Chinese government issued the initiative of Internet + smart energy, indicating that the integration of electricity, thermal energy, and natural gas is an imperative basis to construct the energy internet [7]. With the rapid development of multi-energy systems and energy internet, the coupling of various forms of energy has been growing tighter along the energy production, delivery, and consumption sectors [8].

Demand response is a critical and effective measure to stimulate the demand side resources to interact with renewable generation in the power system [9–11]. According to the reports from Department of Energy (DOE) in the USA, the electricity users can participate in price-based and incentive-based DR programs [12] to shift or reduce load demands. When the secure operation of the power system is jeopardized or there exist large amounts of renewable energy to be consumed, DR can effectively help keep the balance between electricity supply and demand, and accommodate more renewable energy [13, 14]. In most existing research, the DR techniques are applied for a single energy carrier system, e.g., electricity, which are barely feasible when electricity users have some shiftable or curtailable loads [15]. However, inelastic electricity users with only must-run loads

P. Du et al., *Demand Response in Smart Grids*,
https://doi.org/10.1007/978-3-030-19769-8_5

cannot participate in any DR program at all. In addition, due to high discomfort costs, most of the electricity users would operate as must-run loads and are reluctant to interrupt or delay their electricity consumption. Therefore, the single energy carrier system cannot fully utilize the demand side resources to implement DR programs.

The core concept of multi-energy systems brings new insights for demand response. The integration of electricity, thermal energy, natural gas, and other forms of energy enables all the energy users to be active in DR programs [16]. With the complementarity of MESs, the energy users, including must-run loads, can actively participate in DR programs by converting various forms of energy to electricity in peak periods, instead of purchasing electricity from the power system. From the power system perspective, the energy users reduce electricity demands in peak periods. From the users' point of view, their energy consumption is not changed at all, maintaining consumers' comfort. This idea can be extended to the multi-energy systems, where DR becomes a critical measure to improve the economy and reliability of MESs. This novel vision of DR programs is termed as "Integrated Demand Response." By implementing the IDR programs, the MESs, e.g., district heat/cooling, natural gas, biomass, and electric power systems, will back up each other to constitute a more economical and reliable entity. In addition, the response capability of users can be fully exploited without any loss of energy users' comfort.

Up to present, the research on IDR in the MES has been drawing wide attention from the world. In this chapter, the review and prospect of IDR in the MES are conducted to provide a reference for the future investigation. First, the basic concept of IDR and the value analysis are introduced. The research on IDR in the MES is then summarized. The overviews of the engineering projects around the world are introduced. Finally, the key issues and potential research topics on IDR in the MES are proposed.

5.2 Concept of Integrated Demand Response

5.2.1 The Basic Concept of IDR

Integrated demand response is a novel vision of DR programs. Taking advantage of the complementarities of different inertia of multi-energy, IDR is aimed at fully exploiting the DR capabilities of all the users and improving the economic and reliable operations of multi-energy systems. The conventional DR programs are merely focused on the electric power sector. Energy users are encouraged to participate in incentive-based and price-based DR programs [17]. When the power system is jeopardized, incentive payments or discounted rates will be broadcast to energy users for pre-contracted load reductions. Or the energy users with some shiftable loads will actively respond to the change of real-time prices (RTPs) and shift load demands away from peak hours to valley hours. These options usually

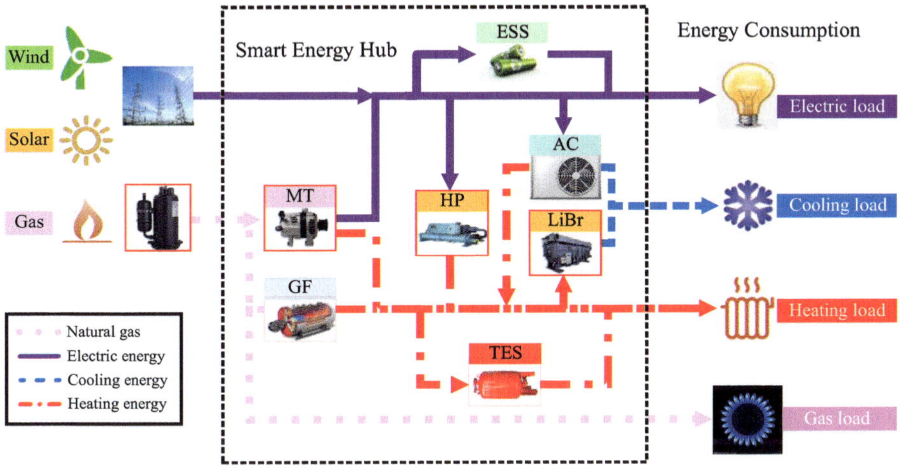

Fig. 5.1 The diagram of integrated demand response

involve a temporary loss of comfort [18–20]. The changes of the electricity consumption behavior cause dissatisfaction and discomfort of energy users. In addition, only a single energy carrier, e.g., electricity, is utilized to implement the DR programs, which may not be applicable for all users. For instance, in power systems, the users with only must-run loads cannot participate in DR programs. Even if some loads are shiftable or curtailable, the DR capability of energy users cannot be fully exploited merely through electricity.

With the development of multi-energy systems, the integration of electricity, thermal energy, natural gas, and other forms of energy enables all the energy users to be active in DR programs. The smart energy hub, as the center of converting various kinds of energy, has played an important role in the MES. By co-optimizing and deploying the synergy of MESs, the flexibility and economy of the energy systems will be significantly improved via the strategic complementarity among different forms of energy. Coupling with different energy carriers, SEHs can economically optimize the input energy flows to satisfy arbitrary energy outputs by converting internal energy resources.

In an IDR program, energy users can not only shift their energy consumption, but also change the source of the consumed energy. Hence, all the energy users, including the must-run loads, can be active to provide DR capabilities for the MESs. A diagram of IDR is shown in Fig. 5.1. In the diagram, the electric energy and natural gas from the power and gas systems are integrated by various devices in the SEH, including energy storage systems (ESSs), micro-turbines (MTs), gas furnaces (GFs), heat pumps (HPs), air conditioning (AC) systems, refrigeration system (e.g., LiBr chiller), thermal energy systems (TESs), etc. In addition, different types of load demands can be satisfied with the optimal operations of the SEH.

One of the major contributions of IDR is to improve the reliability of the MESs. For instance, taking advantage of the inertia of the thermal loads and natural gas

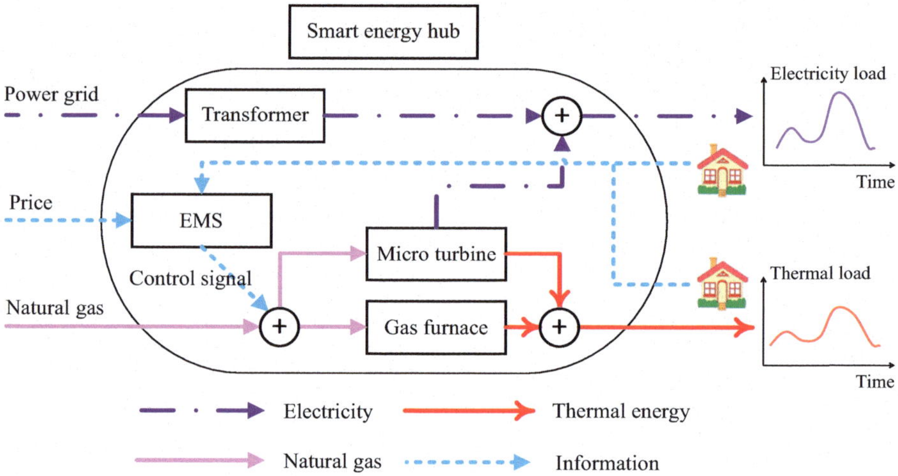

Fig. 5.2 The schematic of a combined gas and electric power system

systems, the reliable operation of the electric power system can be enhanced with the help of IDR. Take the combined gas and electric power system as an example [21–23], the schematic is shown in Fig. 5.2.

The energy inputs of the system are natural gas and electricity from the power system. To satisfy the electricity and heat loads, SEHs can optimize and convert the input energy flows. When the electricity prices are high at peak hours, the natural gas will be mostly supplied for the electricity loads through micro-turbines. Equivalently, the electricity from the power system can be effectively reduced. In addition, due to the thermal inertia of the buildings, the comfortable temperatures will be maintained for a period. When wind output is high at night leading to low electricity prices, the SEHs will control the users' appliances to directly consume electricity from the power system. From the power system's perspective, the energy users reduce electricity demands in peak periods and increase electricity demands in valley hours. From the users' point of view, the electricity consumption is not changed at all and the temperatures can be maintained within the comfortable levels. Therefore, by means of IDR, the DR capability of users can be fully utilized without any loss of energy users' comfort.

5.2.2 *The Value Analysis of IDR*

With the expansion of multi-energy systems towards the demand side, IDR will break down the barriers between electricity and other forms of energy, thereby achieving the deep integration of multiple energy and information streams in the demand side. The concept of IDR fully considers the strategic complementarity of multiple energy consumption, yielding the synergy effects of MESs [24]. The values of IDR in terms of the system operation and users' benefits are analyzed as follows:

5.3 Improve the Economy of Energy Systems

By integrating electricity, thermal energy, natural gas, and other forms of energy, IDR makes it possible for the system operator to maximize the social welfare in a greater optimization space. IDR breaks the barriers among different forms of energy, which enables energy users to flexibly switch the energy sources according to different energy prices. By converting electricity to gas and thermal energy, large amounts of renewable energy can be further accommodated and reused in the future [25]. As a result, the total operational costs can be dramatically reduced. In addition, the strategic complementarity of various energy can improve the utilization of the electric distribution networks, natural gas, and heat supply pipelines. This system-friendly manner will defer the expansion planning of energy distribution systems and save a large amount of investment costs.

5.4 Enhance the Reliability of Energy Systems

Taking advantage of the complementarities of different energy, IDR can help enhance the reliability of energy systems [26]. Different energy systems back up each other to satisfy the load demands of the energy users. When the natural gas is in shortage, the electricity from the bulk power system will be used to meet the requirements of the consumers. Moreover, distinguished from the electric power system where power balance must be kept at any given instant, the energy unbalance is acceptable for the thermal and gas system because of the storage ability. Considering the inertia of natural gas systems, power-to-gas (P2G) technique can be applied to incorporate the surplus of electric power by converting electricity to hydrogen or methane [27, 28]. In addition, the storage ability of the heat/cooling systems can be used to smooth out the fluctuations of electric power and reduce the peak loads [29].

5.5 Exploit the Capability of Demand Side Resources

IDR enables energy users to consume energy in a more flexible manner and fully utilize the capability of DR resources. With the integration of various energy in MESs, SEHs make energy users flexibly change the energy inputs responding to the requirements from the power system or the price signals of different energy. Without any loss of users' comfort, energy users can switch the use of electricity, thermal energy, natural gas, or other forms of energy. In addition, the capability of demand side resources can be fully exploited with the natural storage capability of thermal and gas systems. The surplus of renewable energy can be economically stored in thermal and gas systems and reused later. Hence, the MESs will benefit from the IDR capability in terms of improving the global economy, enhancing the system reliability, etc.

5.5.1 The Techno-Economic Analysis of IDR

Coupling with different energy carriers in the SEHs, the strategic complementarity of various forms of energy makes it possible for all the consumers to actively participate in IDR programs without violating his/her comfort level. In practice, given the prices of electricity, heating, natural gas, and other forms of energy, the consumers would like to manually switch the energy consumption to minimize the energy costs. To automatically implement the IDR programs, an electronic multi-energy system supported by the advanced ICT (Information & Communication Technology) infrastructure needs to be developed. Therefore, it is worth conducting a comprehensive techno-economic analysis of implementing the IDR programs.

One major cost of IDR is to construct the multi-energy facilities and upgrade the sectors of various energy to make the consumers flexibly switch the energy sources. Electricity, cooling, heating, and gas networks need tight interactions through various distributed energy resources (DERs) such as combined heat and power (CHP), HPs, GFs, ACs, refrigeration, etc. Similarly, the electricity, the fuel chain, as well as the transportation sector will be highly coupled with the increase of hybrid vehicles [30]. In the future, it becomes convenient for the consumers to switch natural gas and electric power as the fuel of the vehicles.

With the development of smart grids, advanced metering, communications, control devices, and other information technologies will support the implementation of IDR and become another major cost. The application of ICT for the control of the MESs will lead to the development of an integrated energy and communication system architecture. There have been a number of initiatives across the world making efforts to the use of real-time information with IDR programs to maximize the system operation efficiency while providing the consumers with flexible energy choices; for example, the GridWise and IntelliGrid in the USA and SmartGrids in the EU [31].

The implementation of IDR requires a large amount of investment and operation costs to support the MESs and ICT infrastructures. However, there will be significant benefits brought by the technology upgrading. The benefits of IDR in both economic and environmental perspectives have been widely investigated in the existing research.

IDR is able to break the barriers among different forms of energy, which highly improves the DR ability of the demand side resources [32]. With the consumers' active participation in the IDR programs, the operation efficiency of the energy systems can be further increased and the expansion planning can be effectively deferred. In [33], the economic benefits of IDR are evaluated in the smart buildings based on an integrated electricity-heat-gas model. According to the simulation results in a real UK multi-energy district, the deployment costs of the district can be further minimized by switching different energy consumptions without compromising end-users' comfort. In [34], the economic benefits of the IDR with micro-CHP systems are investigated. The case studies demonstrate that micro-CHP is an energy efficient technology that provides heat and electricity to households. With the optimal scheduling of IDR with micro-CHP, the operation costs for households can be reduced by up to 14%.

The additional value created by IDR is the improvement of the environmental benefits of the energy systems. A better scheduling of different energy sectors can be achieved, leading to pollutant emission reduction. Moreover, with the complementarities of the heat and electric power, the consumers in the IDR programs can provide more ancillary services (e.g., spinning reserve service) for the electric power system instead of the conventional centralized power plants. Hence, the fuel costs and the pollutant emissions can be further decreased. In [35], the concept of IDR is used in the electro-thermal systems to quantify the economic and environmental costs. Taking advantage of the thermal inertia of the thermal systems, the CO_2 emissions of different energy sectors can be further reduced without violating the consumers' comfortable temperature. In [36], the CO_2 emission reduction of a UK dwelling is assessed by applying Stirling engines and fuel cells to satisfy the heat and power demands. That CHP system is predicted to achieve 16% CO_2 emission reductions compared to a non-CHP base case of a condensing boiler.

Therefore, with the development of the MES and ICT, it is beneficial to promote the demand side resources towards integrated demand response.

5.6 IDR Research in the Multi-Energy System

In this section, a systematic literature review on the state of the art of IDR research is conducted. The existing research on IDR can be categorized in different perspectives. The framework is shown in Fig. 5.3.

Spatial scale: element, region, and cross-region level.

Temporal scale: short, medium, and long term.

Energy resource: Electricity, thermal energy, natural gas, biomass, hydrogen, etc.

Research subject: Modeling, operation, market behavior, solution algorithm, etc.

Physical law: Ohm's law, thermodynamics low, fluid mechanics, etc.

In this chapter, the IDR research is reviewed from the perspective of research subject. Specifically, the studies on the modeling, operation strategy, and market behavior of IDR in the MES are focused on.

5.6.1 Modeling

There have been quantities of studies focused on the modeling of the multi-energy devices to implement the IDR programs. Multi-energy conversion can be achieved

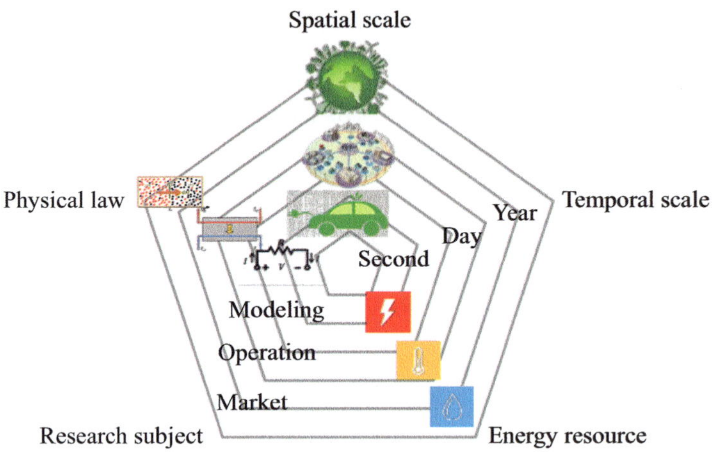

Fig. 5.3 The framework of IDR research in the MES from different perspectives

through the devices, including energy hubs, HPs, CHP systems, fuel cells, P2G systems, etc.

An energy hub is a unit where multiple energy carriers can be converted, conditioned, and stored. An energy hub consumes power at the input ports connected to, e.g., electricity and natural gas infrastructures and provides required energy services such as electricity, heating, cooling, and compressed air at the output ports [37]. Within the energy hub, different technologies are applied to optimally convert the energy flows. In [38], the combined optimization of coupled power flows of different energy infrastructures such as electricity, gas, and district heating is modeled as an energy hub. With that model, optimal energy flow problem is investigated. In [39], the energy hub model is established in a matrix form. The converter model and the storage model are incorporated in the energy hub. A typical example of the energy hub designed in Waterloo, Canada for the supply of commercial loads is introduced. In [40], a new formulation of the energy hub is presented. In that model, the energy system efficiencies, storage losses, operating limits, and other practical constraints are considered. The energy hub model is demonstrated in an example case, where the impacts on the optimal schedule are observed.

Combined heat and power and heat pumps are two efficient systems to satisfy the heat and electric loads. Combined heat and power is the use of a heat engine to generate electricity and useful heat at the same time. In [41], the multi-site CHP planning model is formulated with multiple heat balances. An extended power simplex algorithm is developed to solve the large-scale problem as the number of thermal demand sites increases. In [42], a mixed-integer linear programming model is developed to optimally operate the energy systems in an urban area. In the energy systems, the smart buildings are equipped with CHP and connected by a heat distribution network. A heat pump is a device that transfers heat energy from a source of heat to a heat sink. Heat pumps have very high efficiency, reducing energy consumption up to 72% compared with electric resistance heating [43]. In [44], an enhanced vapor recompression heat pump model is developed to reduce the heat loss and heat pump size. The simulation results show that the developed model yields considerable energy savings compared with the traditional devices. In [45], a hybrid ground source heat pump (GSHP) is modeled, which uses solar thermal collectors as the supplement in smart buildings. The case studies based on the real-world load demands in Milton, Canada show that the costs of the 20-year life cycle are expected to be reduced by 3.7–7.6% compared with the conventional GSHP system.

The energy conversions in the fuel cells and the power-to-gas devices are opposite. In a fuel cell, the chemical energy in the gas fuel is converted to electricity. While in a power-to-gas device, the electrical power is converted to a gas fuel. In [46], the dynamic model of a proton exchange membrane fuel cell is established. The influences of the load demands on the dynamic response of the fuel cells are evaluated. In [47], a physically motivated dynamic model of a proton exchange membrane fuel cell is proposed. The dynamic model is obtained by incorporating the temperature of the membrane. In [28], the power-to-gas process combined with gas seasonal storage operation is modeled. In addition, the physical constraints of natural gas networks are considered in that model. In [27], the power-to-gas process is integrated with the operation of the power system. The potential benefits of the integrated model of power-to-gas are analyzed based on the Great Britain's gas and electric power system. In [48], the Alkaline electrolyzer model and the Catalytic methanation model are introduced in the modeling of a power-to-gas plant. The benefits of the renewable generation accommodation brought by power-to-gas are assessed.

In the existing literature, most studies are focused on the detailed modeling of the multi-energy devices. As the framework of IDR shown in [49], more multi-energy devices can be incorporated in the energy conversion and distribution to construct a comprehensive IDR model. However, the contributions and impacts of the detailed modeling on the system benefits have not been thoroughly investigated, which deserves further research.

5.6.2 Operation Strategy

The multi-energy integration concept of IDR has been widely applied in the existing research on the operation of MESs. The combined heat and power, the integrated gas and power, and other MESs considering the IDR programs have been research highlights.

By implementing IDR, the operational costs of the combined and power systems can be effectively reduced. Moreover, the thermal inertia of the IDR resources, e.g., the smart buildings and the ice storage, can be fully used as potential storage and reserve capabilities of the MESs. In [50], an optimal operation of a CHP-based microgrid is proposed considering energy storage, thermal units, and demand response. In that framework, the IDR resources are treated as virtual generation units. In [51], the potential of a distributed CHP system is quantified to participate in the real-time DR programs. With the energy shifting among heat, cooling, and electric loads, the CHP system is able to respond to DR signals without reducing the energy users' comforts. In [52], a short-term hourly scheduling of industrial and commercial consumers is optimized with the supply of the CHP facilities, power units, and heat-only units. To minimize the operational costs, the IDR program is implemented to meet the heat and electric demands of the consumers, without any curtailed loads. In [53], a day-ahead optimization is carried out for the provision of IDR. Based on the thermal energy storage of hot water and the storage capability of the buildings, the IDR resources can provide more flexibility for the power system.

With the operation of the energy hubs, the consumers can flexibly switch the use of natural gas and electricity. On one hand, the natural gas system and the electric power system back up each other via the IDR programs. On the other hand, the energy demands of the consumers are satisfied without violating the comfort level. In [54], the IDR program is used in the optimal scheduling framework of electricity and natural gas. By the optimization of the energy hubs, the electricity and gas loads of the consumers can be satisfied with minimum system operational costs. In [55], the role of IDR is investigated in the day-ahead scheduling of electric power systems with natural gas systems. The case studies demonstrate the benefits of applying IDR for enhancing the coordination of natural gas and electricity networks. In [23], an integrated model is proposed to dispatch hourly generation and load resources. In that model, the natural gas network is incorporated to supply the gas-fired units. The case studies show that the demand side resources provide a viable option for flexible ramping products when the natural gas is in shortage.

There have been a number of studies incorporating various forms of energy to further achieve the benefits of IDR. In [56], demand response is modeled and integrated to gas, thermal, and electricity infrastructures. Combined with distributed generation, energy storage, CHP, etc., the multi-energy hub is used to serve the consumers' load demands. In [57], an energy hub receives various energy carriers in the input ports, including gas, electricity, and wind power. After conversion, storage, direct connection, and shifting, the energy hub provides electricity, gas, and heat power for the end-use consumers. IDR is a critical measure for the energy hub to minimize the operational costs. In [58], the energy hub is further developed to include local energy technologies such as photovoltaics, biomass, small hydro power, district heating, and building storage. With the integration of various forms of energy, the energy hub allows to lower peaks in electric load demands, thereby reducing the operational costs of the energy systems.

5.6.3 Market Behavior

There have been extensive studies focused on the market behaviors of the IDR resources. Taking advantage of the complementarities of different forms of energy, the capabilities of the IDR resources can be fully exploited. It is worth mentioning that the behaviors of the IDR resources are mainly investigated in the electricity markets in the existing literature. These studies can be generally categorized into two types: (1) the optimal bidding strategies of the IDR resources as price-takers and (2) the game strategies of the IDR resources as price-makers.

As price-takers, given the prices in different markets, the consumers in the IDR programs will strategically allocate the available capacities for energy and ancillary services. In most electricity markets across the world, ancillary services play a critical role in the reliable operation of power systems. In California, for example, a qualified market participator can bid in the joint energy, spinning reserve, and regulation markets. In [59], the IDR programs are implemented in an optimal bidding framework for an MG. In that framework, the consumers' electricity load demands are satisfied with the DERs in the MG while the room temperatures are maintained within the comfort level. By considering the thermal inertia of the buildings, the renewable energy accommodation can be highly improved in the IDR programs. In [29], the IDR capabilities of a CCHP-based MG are investigated in the joint energy and spinning reserve markets. The electric, heating, and cooling demands are satisfied by coordinating various DERs. The simulation results demonstrate that the thermal inertia of the heating/cooling systems can help support the electric power system. In [60], a control framework is proposed for reliable scheduling and provision of frequency reserves by the aggregations of commercial buildings. The IDR capabilities of a Swiss office building are demonstrated by optimizing the load demands of the heating, ventilation, and air conditioning systems. The simulation results show that the IDR capacities can be used for different reserve products in the commercial buildings.

As price-makers, the bidding strategies of different consumers in the IDR programs will be influenced by each other, which further changes the market clearing prices. These problems can be formulated as non-cooperative game models, among which the supply function model has been widely investigated. In [49, 61], the IDR program is formulated for the electricity and natural gas networks. The interactions among the SEHs are modeled as an ordinal potential game. The simulation results show that in addition to load shifting, the consumers in the SEHs can participate in the IDR programs by switching the energy resources during the electricity peak hours. In [62], an energy management framework is proposed for joint operation of CHP and PV prosumers with the price-based IDR. With the MG operator as the leader and the prosumers as the followers, that model is designed as a Stackelberg game.

5.7 IDR Application in the Multi-Energy System

To fully exploit the capability of demand side resources, many countries around the world have been encouraging the research and engineering projects of IDR and impelling the development of multi-energy systems, while achieving the grid-friendly nature of IDR. In this section, the IDR applications in different regions across the world are reviewed.

5.7.1 The USA

In 2001, American Government issued the concept of "Grid 2030" [63], aiming at taking advantage of promising technologies to modernize and expand the Nation's electric power system. Among the new technologies, it is imperative to develop DERs and CHP to integrate various forms of energy in the IDR programs.

The future renewable electric energy delivery and management (FREEDM) system was proposed by the National Science Foundation (NSF) FREEDM System Center for the first time in 2008 [5, 64], presenting an architecture suitable for plug-and-play of distributed renewable energy and distributed energy storage devices. The consumers in the future residential distribution system will be able to flexibly achieve smart energy management with the help of energy routers [65, 66]. By coordinating the home appliances including the air conditioners, the roof-mounted solar panels, heat pumps, etc., the IDR programs can be efficiently developed.

In 2012, an energy startup called Stem developed a battery for commercial buildings to optimally determine the charging and discharging states by predicting the energy prices. Stem's batteries are stripped-down lithium-ion automotive batteries linked to power electronics designed to quickly switch between partially powering a building or charging from the grid [67]. By directly controlling the air conditioners in the commercial buildings, the electric and thermal systems of the commercial buildings are coordinated to reduce the energy costs and achieve the IDR effects. Up to present, dozens of companies have been developing analytics software aimed at such efficiency improvements.

The Stanford Energy System Innovations (SESI) project is a typical IDR program, as a major transformation of the campus district energy system. The transformation is from gas fired combined heat and power with steam distribution to electrically powered combined heat and cooling with hot water distribution. The new heat recovery system is 52% more efficient than the existing cogeneration system on a pure natural gas basis. SESI will cut Stanford's greenhouse gas emissions in half, thereby saving 18% of Stanford's drinking water supply and $300 million operational costs over the next 35 years compared to the existing system.

5.7.2 Europe

In Europe, many countries are prompting the integration of multi-energy in the demand side.

1. Germany.

"E-Energy project" was proposed by Germany Federal Ministry for Economic Affairs and Energy in 2008. Based on the Information Communications Technology (ICT), E-Energy project is aimed at constructing the future multi-energy system with the communication and coordination between the energy systems and home appliances. By integrating the electricity, thermal, and gas energy, the home energy management can be implemented as an IDR program, which will greatly improve the efficiency and reliability of consumers' energy use [68]. With the implementation of E-Energy project, the penetration of distributed energy has reached 50% in Germany by 2015.

2. The UK

The research and projects on the electricity and natural gas integration are highly sponsored by the UK Government [69]. Department of Energy and Climate Change (DECC) in the UK and Innovate UK Corporation jointly funded the project "Energy Systems Catapult" to investigate the integrated energy systems in Birmingham in 2015 [70]. This project is aimed at efficiently satisfying the electric, gas, and district heating demands of energy users, which can also participate in IDR programs.

3. Switzerland

The smart energy hub is a key measure to achieve the IDR programs. The concept of "Energy Hub" was presented for the first time by the Power Systems Laboratory in ETH Zurich in Switzerland [37, 38]. An energy hub exchanges power with the surrounding systems, primary energy sources (e.g., solar, wind, and hydro), loads, and other components via hybrid input and output ports. From a system point of view, an energy hub can be identified as a unit that demonstrates the basic features input, output, conversion, and storage of different energy carriers. The energy hub provides a static and standard model for the integration of different energy forms.

5.7.3 China

In China, it is an important national strategy to accommodate the renewable and sustainable energy and improve the efficiency of energy utilization. In 2015, the Chinese government issued the initiative of Internet + smart energy, indicating that the integration of electricity, thermal energy, and natural gas is an imperative basis to construct the energy internet [7]. Supported by the Chinese government, there have been a few engineering projects on IDR in the recent years.

A high tech industry park has been designed in Zhejiang Province, Jiaxing City since 2013. This project is to build an active distribution network (ADN) with the coordination of roof-mounted photovoltaic and energy storage systems. On one hand, the ADN can satisfy the local load demands in an IDR manner by coordinating different energy resources. On the other hand, the ADN can help enhance the security of the distribution power system. Until 2015, the installed capacity of PV has reached 100 MW. Based on energy management technique, the load demands of the industry park are satisfied by PV systems and the surplus of solar energy is stored in the ESSs. The stored energy can be provided for the distribution system to alleviate the power shortage.

Another microgrid engineering project is located in Qinghai Province, Qilian City with 3.087 MW PV system, 1.2 MWh lithium, and 4 MWh lead-acid batteries. Over 500 home consumers are self-sufficient in the MG project.

Shanghai Tower is a typical MES with the integration of electricity, thermal, and gas energy. Equipped with CHP, air conditioning, ice storage system, ground-sourced heat pump, and gas furnace, the IDR programs can be efficiently implemented. The energy management system in Shanghai Tower can flexibly and automatically optimize the internal energy consumption according to consumer behavior and time-of-use (TOU) prices in Shanghai. By shifting peak loads to the valley hours and optimizing the energy consumption in the building, the energy efficiency can be increased by 22%, and 10 million dollars of operational costs can be saved.

5.8 Key Issues and Potential Researches of IDR

According to the literature review, the research and engineering projects about multi-energy systems and integrated demand response have been widely investigated in the past decades. Most of the existing studies are focused on the optimal operation of MES considering DR. However, there remain more key issues and potential researches to be addressed.

5.9 The Precise Modeling of Multi-Energy Consumption

The existing literature mainly focuses on the generation, distribution, and conversion of different energy systems. The modeling of new-type energy carriers and the applications have been research highlights, including CCHP [71, 72], fuel cells [73], energy storage systems [74], electric vehicles (EVs), natural gas systems [75, 76], ground source heat pumps (GSHPs), ice storage, etc. However, it remains as open questions that (1) how to precisely model the multi-energy consumption sector to reveal the inertia characteristics of IDR and (2) how to evaluate the contributions of the detailed models of multi-energy consumption. An illustrative figure is shown in Fig. 5.4.

In contrast to electric power system where the power consumption must be balanced with the power supply at any instance, the heat/cooling and natural gas consumptions at the demand side are not instant. The processes of the thermal and gas consumptions must obey thermodynamics and hydromechanics principles, respectively. In addition, different facilities have distinct physical characteristics as well as working conditions. For example, the heat transfer rate and efficiency are different when the thermal energy is supplied for the indoor heating system and the water heater. In addition, the consumption flows are different when the natural gas is used for cooking and heating. A precise modeling of multi-energy consumption leads to accurate implementation of IDR. Therefore, it deserves an in-depth investigation in the future research.

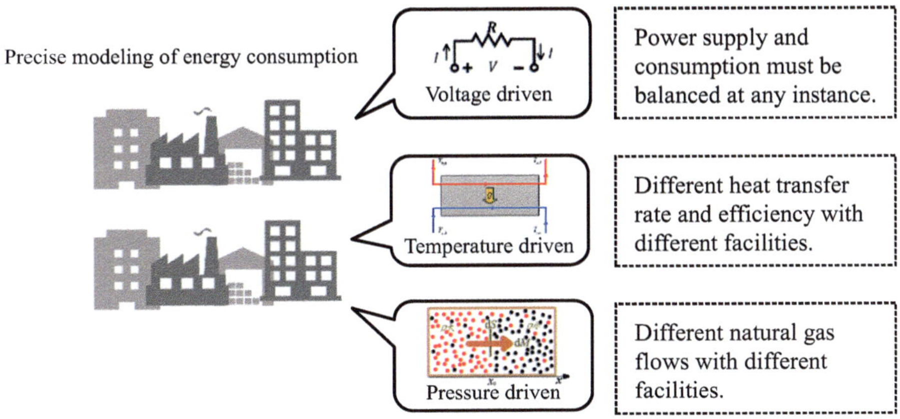

Fig. 5.4 Illustration of the precise modeling of multi-energy consumption

5.10 The Scheduling Strategy for MESs with IDR Considering the Influences of Different Energy Prices

Energy prices are critical incentive signals to lead the energy users to rationally consuming different forms of energy. Hence, it is imperative to incorporate well-designed pricing mechanism in the scheduling strategies for MESs with IDR. According to the incentive compatible principle [77], energy prices should reflect the actual usage of the energy resources among each sector in different energy systems. However, in most existing literature, the energy prices are taken as exogenous conditions, which are predicted as constants or modeled as random variables [78–80]. How to design effective prices according to the response characteristics of IDR resources still needs an in-depth study. For example, it remains as an open question how the consumers will respond to the price changes of different energy in a market-based environment. Another example is how to design effective price signals to alleviate the congestions in the distribution networks and pipelines by taking advantage of the complementarities of different energy.

A scheduling framework of the MESs with IDR is illustrated in Fig. 5.5. On the upper level of the framework, the consumers optimize the consumptions of different energy given the energy prices. On the lower level of the framework, the energy prices are generated from the market clearing based on the scheduling of different consumers. The problem can be formulated as a non-cooperative game model. The model, the solution methodology, and the environmental and economic impacts of the IDR scheduling have not yet been thoroughly investigated.

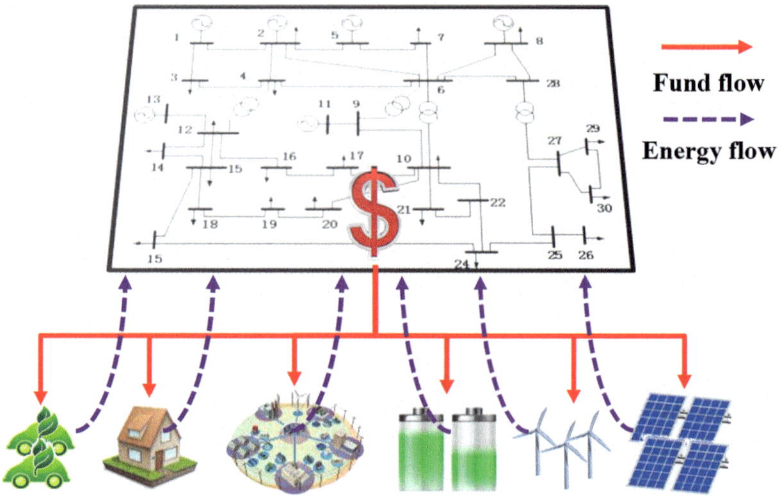

Fig. 5.5 Illustration of scheduling for MESs with IDR considering the influence of different energy prices

5.11 The Data-Driven Consumption Strategy of IDR

In most existing research, the objectives of the energy users are to maximize the individual revenues in the scheduling. However, this may not be the case in the real world. To simulate the energy users' irrational consumption behaviors, Experimental Economics and Consumer Psychology need to be studied by applying the data-driven statistical methods [81].

With the development of advanced metering technology, the real-world energy consumption data can be collected. Then the data-driven technique is adopted to investigate the consumption strategies of IDR [82]. In contrast to the conventional analytic models, the data-driven model and method can reflect the real consumption behaviors of the energy users. The data-driven consumption strategy of IDR is illustrated in Fig. 5.6.

The applications of data-driven IDR are promising and there remain a number of research highlights to be addressed: (1) how to interpret the meanings of the data-driven IDR and make the data-driven IDR more practicable to the system operator; (2) how data-driven IDR can be incorporated in the MES model and influence the operation of the MESs; (3) how data-driven IDR will influence other market participators' decision-making.

5.12 The Market Operation and Mechanism Design Considering IDR

To the best knowledge of the authors, the market mechanism of IDR has not been thoroughly investigated [83]. There remain a number of open questions about the market mechanism of IDR: (1) the optimal bidding strategy of IDR resources in the

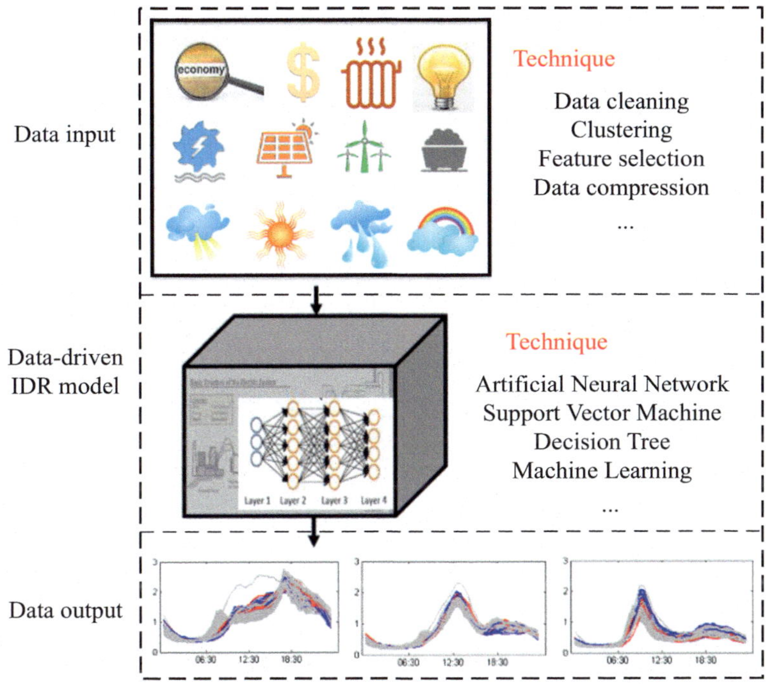

Fig. 5.6 Illustration of the data-driven consumption strategy of IDR

energy and ancillary service markets, which evaluates the capability of IDR as price-takers; (2) the market equilibrium of multiple IDR resources; (3) how to coordinate the bidding strategies of IDR resources in different energy markets along the time horizon, e.g., how to determine the available capacities in the day-ahead electricity markets considering the bilateral contracts in the mid-term natural gas market; (4) how to design a market mechanism to effectively eliminate the asymmetric information in the demand side. These research topics deserve further investigation. A framework of the market research on IDR is shown in Fig. 5.7.

5.13 Conclusion

In the electric power system, the interaction capabilities of demand side resources are limited because of high discomfort costs and must-run loads of energy users. Thus, a novel vision of DR programs termed as "Integrated Demand Response" is proposed in the scope of the multi-energy system. With the integration of various forms of energy, all the energy users are able to actively participate in the DR programs. Taking advantage of the complementarity of MESs, the DR capability can be fully exploited for the MESs without affecting consumers' comfort. Therefore, IDR has become a promising measure to stimulate the demand side resources to interact with

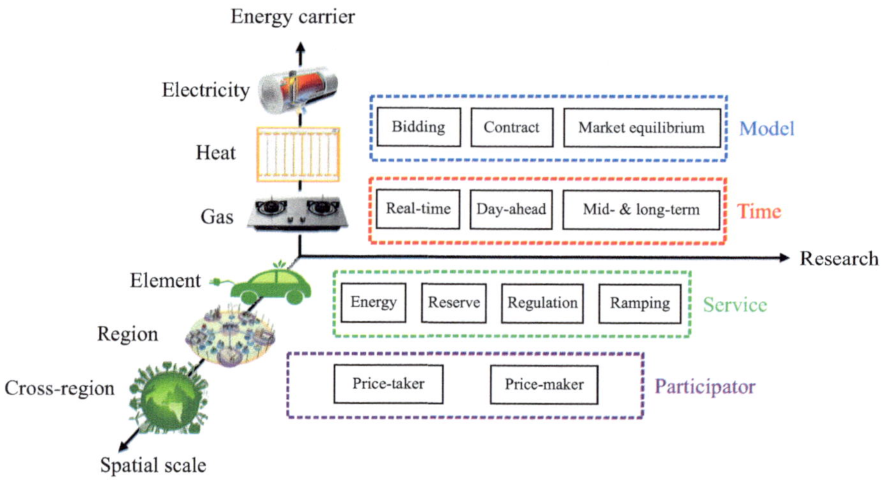

Fig. 5.7 The framework of the market research on IDR

renewable generation in the future. In this chapter, a systematic review on IDR in the MES is conducted for the first time. The basic concept and the value analysis of IDR are firstly introduced. The research on IDR in the MES is then summarized. The overviews of the engineering projects around the world are introduced. Finally, the key issues and potential research topics on IDR in the MES are proposed. The comprehensive review conducted in this chapter will provide reference for the future research and applications on IDR.

References

1. Yu, X. H., & Xue, Y. S. (2016). Smart grids: A cyber-physical systems perspective. *Proceedings of the IEEE, 104*(5), 1058–1070.
2. Ahlgren, L. W. (2012). The dual-fuel strategy: An energy transition plan. *Proceedings of the IEEE, 100*(11), 3001–3052.
3. Glinkowski, M., Hou, J., & Rackliffe, G. (2011). Advances in wind energy technologies in the context of smart grid. *Proceedings of the IEEE, 99*(6), 1083–1097.
4. Kong, F. X., Dong, C. S., Liu, X., et al. (2014). Quantity versus quality: Optimal harvesting wind power for the smart grid. *Proceedings of the IEEE, 102*(11), 1762–1776.
5. Huang, Q. A., Crow, L. M., Heydt, T. G., et al. (2011). The future renewable electric energy delivery and management (FREEDM) system: The energy internet. *Proceedings of the IEEE, 99*(1), 133–148.
6. Rifkin, J. (2011). *The third industrial revolution: How lateral power is transforming energy, the economy and the world*. New York: Palgrave Macmillan.
7. The State Council. *Guidelines of the state council on actively pushing "Internet+" action*. Retrieved June 9, 2016, from http://www.gov.cn/zhengce/content/2015-07/04/content_10002.htm.
8. Krause, T., Andersson, G., Frohlich, K., et al. (2011). Multiple-energy carriers: Modeling of production, delivery and consumption. *Proceedings of the IEEE, 99*(1), 15–27.

9. Biegel, B., Andersen, P., Stoustrup, J., et al. (2016). Sustainable reserve power from demand response and fluctuating production-two Danish demonstrations. *Proceedings of the IEEE, 104* (4), 780–788.
10. Zhong, H., Xie, L., & Xia, Q. (2013). Coupon incentive-based demand response: Theory and case study. *IEEE Transactions on Power Apparatus and Systems, 28*(2), 1266–1276.
11. Nolan, S., & O'Malley, M. (2015). Challenges and barriers to demand response deployment and evaluation. *Applied Energy, 152*, 1–10.
12. US Department of Energy. *Benefits of demand response in electricity markets and recommendations for achieving them: A report to the united states congress pursuant to section 1252 of the energy policy act of 2005*. Retrieved June 9, 2016, from http://www.oe.energy.gov/DocumentsandMedia/congress_1252d.pdf.
13. Siano, P., & Sarno, D. (2016). Assessing the benefits of residential demand response in a real time distribution energy market. *Applied Energy, 161*, 533–551.
14. Finn, P., & Fitzpatrick, C. (2014). Demand side management of industrial electricity consumption: Promoting the use of renewable energy through real-time pricing. *Applied Energy, 113*, 11–21.
15. Chehreghani, M., Hashmi, A., Hasse, H., et al. (2012). Optimal operation of residential EH in smart grids. *IEEE Transactions on Smart Grid, 3*(4), 1755–1766.
16. Yang, H., Xiong, T., Qiu, J., et al. (2016). Optimal operation of DES/CCHP based regional multi-energy prosumer with demand response. *Applied Energy, 167*, 353–365.
17. Zhang, Q., & Li, J. (2012). Demand response in electricity markets: A review. *2012 9th international conference on the European Energy market* (pp. 1–8).
18. Fadlullah, M., Quan, D., Kato, N., et al. (2014). GTES: An optimized game-theoretic demand-side management scheme for smart grid. *IEEE Systems Journal, 8*(2), 588–597.
19. Hong, Y. W., Shahidehpour, M., & Khodayar, M. (2013). Hourly demand response in day-ahead scheduling considering generating unit ramping cost. *IEEE Transactions on Power Apparatus and Systems, 28*(3), 2446–2454.
20. Shao, S., Pipattanasomporn, M., & Rahman, S. (2013). Development of physical-based demand response-enabled residential load models. *IEEE Transactions on Power Apparatus and Systems, 28*(2), 607–614.
21. Alabdulwahab, A., Abusorrah, A., Zhang, X. P., et al. (2015). Coordination of interdependent natural gas and electricity infrastructures for firming the variability of wind energy in stochastic day-ahead scheduling. *IEEE Transactions on Sustainable Energy, 6*(2), 606–615.
22. Zhang, X. P., Shahidehpour, M., Alabdulwahab, A., et al. (2016). Hourly electricity demand response in the stochastic day-ahead scheduling of coordinated electricity and natural gas networks. *IEEE Transactions on Power Apparatus and Systems, 31*(1), 592–601.
23. Zhang, X. P., Che, L., Shahidehpour, M., et al. (2016). Electricity-natural gas operation planning with hourly demand response for deployment of flexible ramp. *IEEE Transactions on Sustainable Energy, 7*(3), 996–1004.
24. Geidl, M., & Andersson, G. (2007). Optimal coupling of energy infrastructures. *2007 IEEE Lausanne Power Tech* (pp. 1398–1403).
25. Behboodi, S., Chassin, D., Crawford, C., et al. (2016). Renewable resources portfolio optimization in the presence of demand response. *Applied Energy, 162*, 139–148.
26. Mazidi, M., Monsef, H., & Siano, P. (2016). Robust day-ahead scheduling of smart distribution networks considering demand response programs. *Applied Energy, 178*, 929–942.
27. Clegg, S., & Mancarella, P. (2015). Integrated modeling and assessment of the operational impact of power-to-gas (P2G) on electrical and gas transmission networks. *IEEE Transactions on Sustainable Energy, 6*(4), 1234–1244.
28. Clegg, S., & Mancarella, P. (2016). Storing renewables in the gas network: Modelling of power-to-gas seasonal storage flexibility in low-carbon power systems. *IET Generation, Transmission & Distribution, 10*(3), 566–575.
29. Wang, J., Zhong, H., Xia, Q., et al. (2016). Optimal joint-dispatch of energy and reserve for CCHP-based microgrids. *IET Generation, Transmission & Distribution*. 2016.
30. Mancarella, P. (2014). MES (multi-energy systems): An overview of concepts and evaluation models. *Energy, 65*, 1–17.

31. Strbac, G. (2008). Demand side management: Benefits and challenges. *Energy Policy, 36*, 4419–4426.
32. Hawkes, A. D., & Leach, M. A. (2007). Cost-effective operating strategy for residential micro-combined heat and power. *Energy, 32*(5), 711–723.
33. Cesena, E. A. M., & Mancarella, P. (2016). Distribution network support from multi-energy demand side response in smart districts. *Innovative Smart Grid Technologies-Asia, 99*, 753–758.
34. Houwing, M., Negenborn, R. R., & De Schutter, B. (2011). Demand response with micro-CHP systems. *Proceedings of the IEEE, 99*(1), 200–213.
35. Good, N., Zhang, L., Navarro-Espinosa, A., & Mancarella, P. (2013). Physical modeling of electro-thermal domestic heating systems with quantification of economic and environmental costs. *Eurocon*, 1164–1171.
36. Peacock, A. D., & Newborough, M. (2005). Impact of micro-CHP systems on domestic sector CO_2 emissions. *Applied Thermal Engineering, 25*(17), 2653–2676.
37. Geidl, M., Koeppel, G., Favre-Perrod, P., et al. (2007). Energy hubs for the future. *IEEE Power and Energy Magazine, 5*(1), 24–30.
38. Geidl, M., & Andersson, G. (2007). Optimal power flow of multiple energy carriers. *IEEE Transactions on Power Systems, 22*(1), 145–155.
39. Parisio, A., Del Vecchio, C., & Vaccaro, A. (2012). A robust optimization approach to energy hub management. *International Journal of Electrical Power & Energy Systems, 42*(1), 98–104.
40. Evins, R., Orehounig, K., Dorer, V., et al. (2014). New formulations of the 'energy hub' model to address operational constraints. *Energy, 73*, 387–398.
41. Rong, A., Hakonen, H., & Lahdelma, R. (2006). An efficient linear model and optimisation algorithm for multi-site combined heat and power production. *European Journal of Operational Research, 168*(2), 612–632.
42. Bracco, S., Dentici, G., & Siri, S. (2013). Economic and environmental optimization model for the design and the operation of a combined heat and power distributed generation system in an urban area. *Energy, 55*, 1014–1024.
43. Energy Government. *Choosing and installing geothermal heat pumps*. Retrieved March 13, 2017, from https://energy.gov/energysaver/choosing-and-installing-geothermal-heat-pumps
44. Waheed, M. A., Oni, A. O., Adejuyigbe, S. B., et al. (2014). Performance enhancement of vapor recompression heat pump. *Applied Energy, 114*, 69–79.
45. Rad, F. M., Fung, A. S., & Leong, W. H. (2013). Feasibility of combined solar thermal and ground source heat pump systems in cold climate, Canada. *Energy and Buildings, 61*, 224–232.
46. Corrêa, J. M., Farret, F. A., Canha, L. N., et al. (2004). An electrochemical-based fuel-cell model suitable for electrical engineering automation approach. *IEEE Transactions on Industrial Electronics, 51*(5), 1103–1112.
47. Pasricha, S., & Shaw, S. R. (2006). A dynamic PEM fuel cell model. *IEEE Transactions on Energy Conversion, 21*(2), 484–490.
48. Huang, J. H., Zhou, H. S., Wu, Q. H., et al. (2016). Assessment of an integrated energy system embedded with power-to-gas plant. *Innovative Smart Grid Technologies-Asia, 77*, 196–201.
49. Sheikhi, A., Rayati, M., Bahrami, S., et al. (2015). Integrated demand side management game in smart energy hubs. *IEEE Transactions on Smart Grid, 6*(2), 675–683.
50. Aghaei, J., & Alizadeh, M. I. (2013). Multi-objective self-scheduling of CHP (combined heat and power)-based microgrids considering demand response programs and ESSs (energy storage systems). *Energy, 55*, 1044–1054.
51. Mancarella, P., & Chicco, G. (2013). Real-time demand response from energy shifting in distributed multi-generation. *IEEE Transactions on Smart Grid, 4*(4), 1928–1938.
52. Alipour, M., Zare, K., & Mohammadi-Ivatloo, B. (2014). Short-term scheduling of combined heat and power generation units in the presence of demand response programs. *Energy, 71*, 289–301.
53. Good, N., Karangelos, E., Navarro-Espinosa, A., et al. (2015). Optimization under uncertainty of thermal storage-based flexible demand response with quantification of residential users' discomfort. *IEEE Transactions on Smart Grid, 6*(5), 2333–2342.

54. Sheikhi, A., Bahrami, S., & Ranjbar, A. M. (2015). An autonomous demand response program for electricity and natural gas networks in smart energy hubs. *Energy, 89*, 490–499.
55. Zhang, X., Shahidehpour, M., Alabdulwahab, A., et al. (2016). Hourly electricity demand response in the stochastic day-ahead scheduling of coordinated electricity and natural gas networks. *IEEE Transactions on Power Systems, 31*(1), 592–601.
56. Pazouki, S., Haghifam, M. R., & Olamaei, J. (2013). Economical scheduling of multi carrier energy systems integrating renewable, energy storage and demand response under energy hub approach. *Smart Grid Conference, 67*, 80–84.
57. Pazouki, S., Haghifam, M. R., & Moser, A. (2014). Uncertainty modeling in optimal operation of energy hub in presence of wind, storage and demand response. *International Journal of Electrical Power & Energy Systems, 61*, 335–345.
58. Orehounig, K., Evins, R., & Dorer, V. (2015). Integration of decentralized energy systems in neighbourhoods using the energy hub approach. *Applied Energy, 154*, 277–289.
59. Nguyen, D. T., & Le, L. B. (2014). Optimal bidding strategy for microgrids considering renewable energy and building thermal dynamics. *IEEE Transactions on Smart Grid, 5*(4), 1608–1620.
60. Vrettos, E., Oldewurtel, F., & Andersson, G. (2016). Robust energy-constrained frequency reserves from aggregations of commercial buildings. *IEEE Transactions on Power Systems, 31* (6), 4272–4285.
61. Bahrami, S., & Sheikhi, A. (2016). From demand response in smart grid toward integrated demand response in smart energy hub. *IEEE Transactions on Smart Grid, 7*(2), 650–658.
62. Ma, L., Zhang, J., Tushar, W., & Yuen, C. (2016). Energy management for joint operation of CHP and PV prosumers inside a grid-connected microgrid: A game theoretic approach. *IEEE Transactions on Industrial Informatics, 12*(5), 1930–1942.
63. *DOE of United States.Grid 2030: A national vision for electricity's second 100 years*. Retrieved June 9, 2016, from https://www.ferc.gov/eventcalendar/Files/20050608125055-grid-2030.pdf.
64. Huang, A. (2010). FREEDM system-a vision for the future grid. *IEEE PES General Meeting IEEE, 18*, 1–4.
65. Xu, Y., Zhang, J., Wang, W., et al. (2011). Energy router: Architectures and functionalities toward Energy internet. Smart grid communications. *2011 IEEE international conference on IEEE, 16*, 31–36.
66. Zhang, J., Wang, W., & Bhattacharya, S. (2012). Architecture of solid state transformer-based energy router and models of energy traffic. *IEEE PES Innovative Smart Grid Technologies IEEE, 2012*, 1–8.
67. MIT Technology Review. *A startup's smart batteries reduce buildings' electric bills*. Retrieved June 9, 2016, from https://www.technologyreview.com/s/506776/a-startups-smart-batteries-reduce-buildings-electric-bills/.
68. Vermesan, O., Blystad, L. C., Zafalon, R., et al. (2011). *Internet of energy–Connecting energy anywhere anytime* (Advanced microsystems for automotive applications) (pp. 33–48). Berlin: Springer.
69. Qadrdan, M., Chaudry, M., Wu, J., et al. (2010). Impact of a large penetration of wind generation on the GB gas network. *Energy Policy, 38*(10), 5684–5695.
70. CATAPULT. *Energy Systems Catapult*. Retrieved June 9, 2016, from http://catapult.org.uk/energy-systems-catapult/.
71. Zhang, X., Karady, G. G., & Ariaratnam, S. T. (2014). Optimal allocation of CHP-based distributed generation on urban energy distribution networks. *IEEE Transactions on Sustainable Energy, 5*(1), 246–253.
72. Chen, X., Xia, Q., Kang, C., et al. (2012). A rural heat load direct control model for wind power integration in China. *2012 IEEE Power and Energy Society General Meeting* (pp. 1–6).
73. Ellis, M. W., Von Spakovsky, M. R., & Nelson, D. J. (2001). Fuel cell systems: Efficient, flexible energy conversion for the 21st century. *Proceedings of the IEEE, 89*(12), 1808–1818.
74. Lemofouet, S., & Rufer, A. (2005). Hybrid energy storage systems based on compressed air and supercapacitors with maximum efficiency point tracking. *Power electronics and applications, 2005 European conference on*. IEEE 2005: 10.

75. Mukherjee, U., Walker, S., Maroufmashat, A., et al. (2016). Power-to-gas to meet transportation demand while providing ancillary services to the electrical grid. *Smart Energy Grid Engineering, IEEE, 89*, 221–225.
76. Liu, C., Shahidehpour, M., Fu, Y., et al. (2009). Security-constrained unit commitment with natural gas transmission constraints. *IEEE Transactions on Power Systems, 24*(3), 1523–1536.
77. Tushar, W., Chai, B., Yuen, C., et al. (2016). Energy storage sharing in smart grid: A modified auction-based approach. *IEEE Transactions on Smart Grid, 7*(3), 1462–1475.
78. Abbasi, A. R., & Seifi, A. R. (2014). Simultaneous integrated stochastic electrical and thermal energy expansion planning. *IET Generation, Transmission & Distribution, 8*(6), 1017–1027.
79. Yao, W., Zhao, J., Wen, F., et al. (2014). A multi-objective collaborative planning strategy for integrated power distribution and electric vehicle charging systems. *IEEE Transactions on Power Systems, 29*(4), 1811–1821.
80. Salimi, M., Ghasemi, H., Adelpour, M., et al. (2015). Optimal planning of energy hubs in interconnected energy systems: A case study for natural gas and electricity. *IET Generation, Transmission & Distribution, 9*(8), 695–707.
81. Kirschen, D. S., Strbac, G., Cumperayot, P., et al. (2000). Factoring the elasticity of demand in electricity prices. *IEEE Transactions on Power Systems, 15*(2), 612–617.
82. Solanki, B. V., Raghurajan, A., Bhattacharya, K., et al. (2015). Including smart loads for optimal demand response in integrated Energy Management Systems for Isolated Microgrids. *IEEE Transactions on Smart Grid, 99*, 1–10.
83. Wang, Q., Zhang, C., Ding, Y., et al. (2015). Review of real-time electricity markets for integrating distributed energy resources and demand response. *Applied Energy, 138*, 695–706.

Future Reading

1. Neyestani, N., Damavandi, M. Y., Shafie-khah, M., et al. (2015). Uncertainty characterization of carrier-based demand response in smart multi-energy systems. *2015 IEEE 5th international conference on power engineering, Energy and electrical drives IEEE* (pp. 366–371).
2. Arteconi, A., Patteeuw, D., Bruninx, K., et al. (2016). Active demand response with electric heating systems: Impact of market penetration. *Applied Energy, 177*, 636–648.
3. Pazouki, S., Haghifam, M. R., & Olamaei, J. (2013). Economical scheduling of multi carrier energy systems integrating renewable, Energy storage and demand response under Energy hub approach. *Smart Grid Conference IEEE, 12*, 80–84.
4. Gitizadeh, M., Farhadi, S., & Safarloo, S. (2014). Multi-objective energy management of CHP-based microgrid considering demand response programs. *Smart Grid Conference IEEE, 15*, 1–7.
5. Zhang, X., Shahidehpour, M., Alabdulwahab, A., et al. (2015). Optimal expansion planning of energy hub with multiple energy infrastructures. *IEEE Transactions on Smart Grid, 6*(5), 2302–2311.
6. Arteconi, A., Hewitt, N. J., & Polonara, F. (2012). State of the art of thermal storage for demand-side management. *Applied Energy, 93*, 371–389.
7. Lopes, J. A. P., Soares, F. J., & Almeida, P. M. R. (2011). Integration of electric vehicles in the electric power system. *Proceedings of the IEEE, 99*(1), 168–183.

Chapter 6
Coupon Incentive-Based Demand Response

6.1 Introduction

Demand response (DR) offers one of the most promising opportunities for engineering sustainable electricity services in the future. The US Department of Energy defines DR as "a tariff or program established to motivate changes in electric use by end-use customers in response to changes in the price of electricity over time, or to give incentive payments designed to induce lower electricity use at times of high market prices or when grid reliability is jeopardized" [1]. The US Federal Energy Regulatory Commission (FERC) estimates that the contribution from existing DR resources in the USA is around 41,000 MW, which was 5.8% of the 2008 summer peak demand [2]. As the penetration of intermittent generation increases, well designed and utilized DR could lead to great system-wide benefits in terms of improving power system security and economic efficiency [3].

In the existing literature, DR has been categorized into price-based DR and incentive-based DR programs [1].

Various types of price-based DR have been proposed, such as time-of-use pricing (TOU), critical peak pricing (CPP), peak load pricing (PLP), and real-time pricing (RTP) [4–6]. Under these pricing schemes, the fluctuating wholesale electricity price risks are imposed upon retail customers in a mandatory manner. Recent studies show that the demand response in the CPP schema is relatively stable [7]. Hence, the CPP should be helpful in reducing the peak load. A pioneering program that exposed retail customers to hourly RTP was successfully tested in Chicago in 2003 and expanded in 2007 [8, 9]. However, most retail customers are risk-averse and not used to making decisions about electricity consumption on a daily or hourly basis [1]. Furthermore, equity problems might arise from time-varying retail rate schemes, such as the day shift versus the night shift [10]. TOU pricing is also an involuntary way for the electricity consumption of consumers to be adjusted among different time scales in accordance with the cost of electricity [11]. In addition, concerns have been raised that RTP may increase the price volatility [12]. While theoretically

P. Du et al., *Demand Response in Smart Grids*,
https://doi.org/10.1007/978-3-030-19769-8_6

attractive, time-varying retail rate schemes still face obstacles in many regions when it comes to large-scale deployment.

A variety of incentive-based DR programs have also been proposed. Incentive-based DR programs are responsible for 93% of peak load reduction in the USA [13]. Interruptible load contract (ILC) is a popular approach to inducing demand reduction that has been adopted by utilities since the 1980s [14–16]. While consumers receive financial incentives provided by the utilities [17], recent field experience has shown a potential reluctance among consumers of such programs due to the inconvenience brought on by the mandatory interruption of electricity services [18]. In [19], an optimal power flow (OPF)-based framework is proposed to determine the incentive rates of an interruptible tariff mechanism on an hourly basis. In [20], an incentive mechanism is designed for lowest cost aggregate energy demand reduction, in which the utility achieves minimal total operating cost. Peak time rebate (PTR) is another type of incentive-based DR program. An experiment involving 123 retail customers in Anaheim, California was conducted in 2005 [21]. Another PTR pilot study was conducted in Ontario, Canada in 2007 [22]. However, the rebates paid to consumers in both pilot programs have a predetermined fixed value that does not reflect the actual supply-demand status under different sets of operating conditions.

In this chapter, a type of DR program, referred to as coupon incentive-based demand response (CIDR), is introduced; this program attempts to induce flexibility in retail customers (such as small/medium size commercial, industrial, and residential customers) on a voluntary basis [23]. In the USA, there are two types of utilities: private and public [24]. Utilities have a certain rate of investment return on an annual or multi-year basis. In addition to the public utilities, there are also many investor-owned utilities (IOUs), third-party retail aggregators that are profit-seeking. For short-term operations, having this DR will benefit both public and private utilities by reducing the need for peaking units. The CIDR discussed in this chapter applies to the overlapping portions of IOUs and profit-seeking LSEs. Given the fact that the majority of retail customers still enjoy flat rate electricity, and that LSEs purchase time-varying electricity from the wholesale market, LSEs will have a motive to induce the consumers' inherent elasticity [25] by use of voluntary coupon incentives when the wholesale real-time electricity price exceeds the fixed retail rate. Statistics based on the historical price data obtained from the Electric Reliability Council of Texas (ERCOT) show that, in the summer peak load month, the wholesale RTP exceeds the retail rate more than 8% of the time [26]. Most independent system operators (ISOs)/regional transmission organizations (RTOs) in the USA implement both the day-ahead market and the real-time market; this is well known as the two-settlement system [27]. The energy cleared in real-time markets is around 2%~8% [28, 29]. Although this percentage looks small, it represents a huge amount of business in the electricity market. Inspired by the overbooking strategy of the airline industry, LSEs, in order to hedge the temporal risks associated with wholesale real-time price volatility, could induce retail customers to reduce nonessential electricity use voluntarily at times when real-time price spikes are likely to occur by providing the consumers with a financial reward for their response behaviors [30]. Typically, consumers are willing to reduce their power consumption as long as the coupon revenue they receive exceeds the benefit they would achieve if they used the power they choose not to consume [31]. With the emergence of pervasive mobile communication capabilities and smart grid technologies such as smart meters and the home area network (HAN) [32–34], a much tighter information feedback loop among the ISO/RTO, LSEs, and retail customers has become feasible. Consumers' response to such coupon incentives could be realized in close to real time (e.g., min). Due to the fact that consumers are much less likely to increase their electricity consumption in the case of a short-term price drop than they are to reduce their electricity consumption in the case of a price increase [35], this chapter only discusses the latter.

There are two ways to implement voluntary incentives to retail customers in CIDR. One way is to maximize the profit from the LSEs' perspective, whereas the other way is to maximize the social welfare from the ISO/RTOs' perspective. This chapter focuses on the first type of CIDR.

The CIDR scheme offers several benefits as a complementary approach to implementing demand response at the retail level.

First, from the LSEs' point of view, when electricity price spikes are expected on the wholesale market, the LSEs could offer coupon incentives to induce demand reduction instead of suffering the high price spikes. Even if no consumer responds to the coupon incentives, it will not be any worse than the flat retail rate status quo. The study in this chapter shows that even with moderate customer participation, the algorithm can still bring major benefits to the LSE.

Second, from the consumers' point of view, CIDR is a voluntary program in which consumers have the right to choose whether to participate. Compared with a price-based DR that imposes fluctuating wholesale electricity prices upon consumers, CIDR maintains a flat retail rate on the consumers' side and offers financial rebates to the DR providers. Hence, it is much more acceptable to retail customers.

Third, from the perspective of social welfare, based on the assumption of consumer rationality in microeconomic theory, effective demand reduction can be achieved and the utilization of expensive generation resources can be decreased. CIDR is shown to achieve almost the same social welfare as that in the RTP scheme. Future study will investigate the performance of CIDR without the assumption of consumer rationality.

Last but not least, CIDR is compatible with the existing flat rate structure in the retail electricity market. It can be easily deployed in the near future.

The remainder of this chapter is organized as follows: in Sect. 6.2, the CIDR model is formulated and the optimal coupon pricing algorithm is proposed. In Sect. 6.3, the proposed CIDR scheme is analyzed theoretically in terms of social welfare, consumer surplus, LSE profit, the robustness of the retail rate, and readiness for implementation. The pros and cons of CIDR are also discussed. In Sect. 6.4, a numerical case study based on realistic supply and demand data obtained from ERCOT is performed to evaluate the benefits of the proposed CIDR. In Sect. 6.5, the major findings are summarized and future work is suggested.

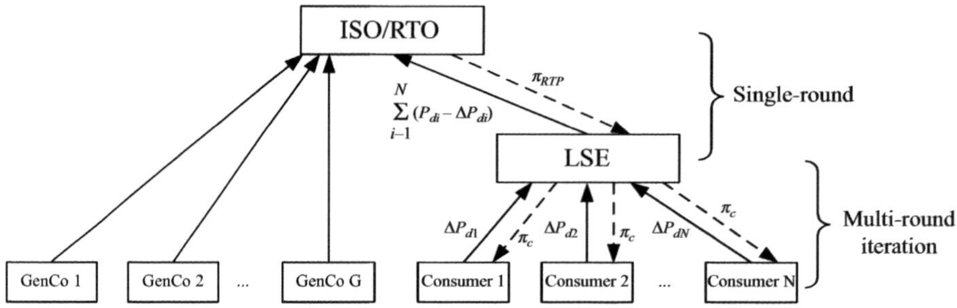

Fig. 6.1 Three-layer information exchange structure for CIDR implementation

6.2 The CIDR Formulation

The CIDR scheme is formulated with a single-stage model of the real-time electricity market. CIDR in a multi-stage real-time electricity market will be investigated in our future work. The coupon incentives are assumed to be uniform for all retail customers.

6.2.1 *Information Exchange Scheme*

In Fig. 6.1, a three-layer information exchange scheme among ISO/RTO, LSEs, and consumers is shown.

The targeted retail customers face a flat electricity rate. In real-time operations, the ISO/RTO publishes expected significant changes in system conditions and/or adverse weather conditions several hours ahead of an operating interval [37]. If LSEs can anticipate the occurrence of price spikes with the updated information, they can issue coupon incentives to retail customers in order to offset the potential financial losses.

Since LSEs cannot perfectly estimate the consumer response to coupon incentives (similar to the issuing of vouchers in the case of airline industry overbooking), several rounds of iterations could be executed before LSEs submit their updated real-time demand bids to the ISO/RTO.

At the highest layer, the ISO/RTO collects the updated real-time demand bids (after incorporating the impact of the coupons on committed retail customers) from the LSE side and the supply offers from the generation side. Then the ISO/RTO clears the market using security constrained economic dispatch (SCED) and the wholesale RTP is determined. This wholesale RTP will be imposed upon LSEs in the settlement process.

At the middle layer, (1) LSEs mimic the ISO/RTO market clearing function using historical and/or estimated supply and demand data, and determine the expected wholesale RTP; (2) they adjust the coupon price according to the expected wholesale RTP and collect demand reduction offers from retail customers. This is a multi-round procedure, since the consumers' willingness to reduce the electricity consumption could be time-varying. For instance, whether the consumer would shut down the air conditioner depends on the near-real-time situation, such as the current room temperature. The CIDR provides such a mechanism that the consumers can flexibly decide whether they would like to reduce the electricity consumption based on the near-real-time situation. LSEs could increase the coupon price gradually until the profit doesn't increase (or equivalently the financial loss doesn't decrease) from the previous iteration.

At the lowest layer, the individual consumer makes his/her own decision when he/she is offered the given coupon price, and then submits his/her demand reduction offer to the LSEs. There are two options to implement the demand reduction offer. One way is to submit the reduction quantity iteratively. The other way is to submit the demand reduction bid (reduction schedule corresponding to various coupon prices) only once. The consumer's willingness to reduce the power consumption is implied in the demand reduction bid. This chapter focuses on the first option.

6.2.2 Timeline of CIDR

The timeline of implementing CIDR in practical electricity markets is illustrated in Fig. 6.2.

From the operational perspective, the ISO/RTO would publish expected significant changes in system conditions and/or adverse weather conditions for the remaining hours of the current operating day [37]. With this updated physical operating alert message, LSEs could anticipate a wholesale price spike based on historical and/or estimated wholesale supply and demand data. They could then broadcast coupon incentive information to consumers via smart grid communication networks.

The pre-operating interval of real-time markets varies in different regions. In the case of ERCOT, it is 60 min [37], whereas it is 75 min in California ISO [38]. The operating interval usually consists of consecutive 5-min SCED procedures and 15-min settlement intervals.

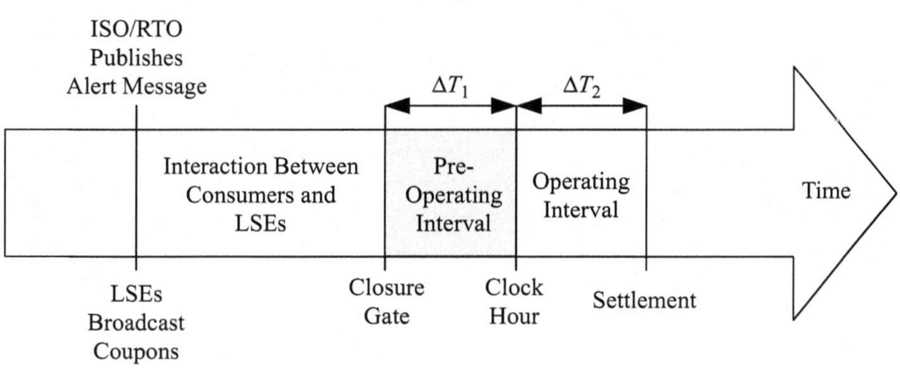

Fig. 6.2 Timeline for CIDR implementation

Prior to the gate closure of real-time markets, LSEs and consumers can interact iteratively (update the coupon price and corresponding demand reduction offers) until LSE profit no longer increases (or equivalently LSE financial loss no longer decreases) from the previous iteration. Facilitated by near-real-time information networking capabilities and automatic energy consumption controls, there will be enough time (approximately 1 h) for retail customers to make their own decisions and adjust the energy usage schedule.

With regards to financial settlement, the settlement procedure will start after the operating interval. The entire demand response of participating consumers will be recorded in the interval power consumption database. The measurement and verification rely on the interval smart meters and the historical database of consumer electricity consumption. The demand reduction is defined as the difference between the baseline and the actual consumption during CIDR event intervals. The verification also relies on the baseline calculation, which will be discussed in Sect. 6.2.3. At the end of a settlement circle (e.g., 1 month), the coupon payment will show up as an independent item on consumers' electricity bills.

6.2.3 Choice of Consumer Baselines

The consumer baseline is the power consumption level from which the demand reduction quantity is calculated. Incentive-based DR programs in electricity markets depend critically on the choice of consumer baselines [36].

In the implementation of the CIDR, there are several ways to obtain the consumer baselines. The consumer baseline can be chosen based on the historical power consumption data of the consumers within a specific time window. This is due to the fact that electricity consumption exhibits strong cyclic patterns over time [39]. The consumer baseline may also be selected based on their power consumption between the instant when the LSE broadcasts the coupon information and the operating interval.

Reference [36] reviews alternative approaches for establishing consumer baselines. In general, there are two types of approaches: administrative and contractual approaches. For retail customers, it would be easier and more appropriate to implement the administrative approach. In terms of implementation, the Southern California Edison utility uses a method called "10-Day Average Baseline" [40]. The baseline is calculated based on the consumption data of the 10 most recent similar days. Reference [41] reviews baseline calculation methods in great detail. For a retail customer, it is very difficult to game across a 10-day period. In addition, the CIDR is activated based on near-real-time system conditions. It is also difficult to foresee when LSEs will offer the coupon incentives. In light of both these things, the likelihood of a retail customer manipulating baselines is small.

The LSEs may choose one of the above methods to compute consumers' baseline. It is out of the scope of this chapter to discuss the pros and cons of various baseline methods.

6.2.4 ISO/RTO

An ISO/RTO is a not-for-profit organization in standard market design that has the objective of maximizing social welfare. Given the fact that the majority of retail customers still enjoy a flat rate structure, LSEs submit expected (inelastic) demand to the ISO/RTO on the real-time market. The objective of the ISO/RTO in real-time operations becomes least cost economic dispatch [37].

The variables used in the problem formulation are defined as follows:

G	Number of generators
N	Number of consumers
P_{gi}	Output of generator i
${P_{gi}}^{\max}$	Maximum output of generator i
${P_{gi}}^{\min}$	Minimum output of generator i
P_d	Total demand in the wholesale market
ΔP_d	Total demand response in the wholesale market
ΔP_{di}	Demand reduction of consumer i
$P_{di,0}$	Baseline of consumer i
P_{di}	Power consumption of consumer i
π_{RTP}	Real-time price in the wholesale market
π_{RR}	Retail rate of electricity
$\pi_{WTP,i}$	Willingness to pay of consumer i
π_c	Coupon price in the retail market
y_i	Utility of consumer i
$b_{i,k}$	Intercept of segment of consumer i's utility function
$C_i(\cdot)$	Cost function of generator i
$g_i(\cdot)$	Response function of consumer i
$U_i(\cdot)$	Utility function of consumer i

Therefore, the formulation is stated as follows:

Objective function:

$$\min_{P_{gi}} \sum_{i=1}^{G} C_i(P_{gi}) \tag{6.1}$$

subject to

$$\sum_{i=1}^{G} P_{gi} = \sum_{j=1}^{N} \left(P_{dj} - \Delta P_{dj}\right) \tag{6.2}$$

$$P_{gi}^{\min} \leq P_{gi} \leq P_{gi}^{\max}, i = 1,2, \ldots, G \tag{6.3}$$

P_{dj} is the consumer baseline of retail customer j. $(P_{dj} - \Delta P_{dj})$ is the updated inelastic demand submitted by LSEs to the ISO/RTO.

6.2.5 LSE

As profit-seeking participants, LSEs' objectives are to maximize their expected profits.

Objective function:

$$\max_{\pi_c} E[\pi_{RR}(P_d - \Delta P_d) - \pi_{RTP}(P_d - \Delta P_d) - \pi_c \Delta P_d] \tag{6.4}$$

subject to

$$\Delta P_d = \sum_{i=1}^{N} \Delta P_{di} \tag{6.5}$$

$$\Delta P_{di} = g_i(\pi_C) \tag{6.6}$$

The first term in the objective function is the retail revenue from consumers, the second term is the purchasing cost from the wholesale real-time market, and the last term is the coupon payment to consumers.

The function $g_i(\pi_C)$ reflects the response behavior of consumer i when he/she is offered the coupon price π_C. Under the assumption of linear demand curves, $g_i(\pi_C)$ can be explicitly expressed as

$$\Delta P_{di} = k_i \pi_C \tag{6.7}$$

where k_i is the inherent demand elasticity of consumer i. However, we noted that, in general, $g_i(\cdot)$ may not be expressed in closed forms and can be obtained through the decision-making process mentioned below in (6.9)–(6.10).

6.2.6 Consumer

In this chapter, retail customers are assumed to be rational participants. The consumers' objectives are to maximize their own surplus. Based on the demand curve with a staircase shape, the utility function of consumer is a concave -segment piecewise linear function [43]:

$$U_i(P_{di}) = \begin{cases} \pi^1_{WTP,i} P_{di} + b_{i,1}, & 0 \le P_{di} \le P^1_{di} \\ \pi^2_{WTP,i} P_{di} + b_{i,2}, & P^1_{di} \le P_{di} \le P^2_{di} \\ \vdots & \vdots \\ \pi^m_{WTP,i} P_{di} + b_{i,m}, & P^{m-1}_{di} \le P_{di} \le P^m_{di} \end{cases} \tag{6.8}$$

The sequence of points $\left(\pi_{WTP,i}^{k}, P_{di}^{k}\right)$, $k = 1,2, \ldots, m$ comes from the demand curve. $b_{i,k}$, $k = 1, 2, \ldots m$ is the intercept of segment of the utility function.

Objective function:

$$\max_{P_{di}} y_i - \pi_{RR} P_{di} + \pi_C (P_{di,0} - P_{di}) \tag{6.9}$$

subject to

$$y_i \leq \pi_{WTP,i}^{k} P_{di} + b_{i,k}, k = 1,2, \ldots, m \tag{6.10}$$

$P_{di,0}$ is the baseline of consumer i. y_i is the auxiliary variable denoting the utility of consumer i. As long as the sum of electricity bill savings and the coupon incentives from LSEs outweighs the lost benefit, consumers are assumed to respond to the coupon incentives. The higher the coupon price is, the greater the number of consumers who will respond to the incentives. $\Delta P_{di} = P_{di,0} - P_{di}$ will then be submitted to LSEs for them to make the decision in (6.4)–(6.6).

6.2.7 Adjustment of the Coupon Price

A multi-iteration coupon price adjustment procedure is presented in Fig. 6.3.

1. When a wholesale price spike is expected, LSEs mimic the ISO/RTO's market clearing function using historical and/or estimated supply and demand data, and determine the expected wholesale RTP $\pi_{t,0}$ assuming that there is no DR. $\pi_{t,0}$ is usually greater than the flat retail rate; otherwise, LSEs would have no incentive to implement the CIDR scheme.
2. LSEs set the initial coupon price at $\pi_{t,c,0} = \lambda(\pi_{t,0} - \pi_{RR})$, $0 < \lambda < 1$. The iteration indicator m is set to be 1.
3. LSEs broadcast coupon incentive information to consumers via near-real-time information networks (e.g., social networks).
4. Consumers respond to the coupon incentives and submit their binding demand reduction offers $\Delta P_{d,i,t,m}$ to LSEs.
5. LSEs mimic the ISO/RTO to clear the market and determine a new expected wholesale RTP $\pi_{t,m}$ with the DR submitted by consumers.

The convergence criterion is that the LSE profit no longer increases (or equivalently LSE financial loss no longer decreases) from the previous iteration.

The consumers can decide whether to take the coupons. Hence, consumer decisions play a very important role in this process. In terms of implementation, consumer response can be in some ways automated. For commercial & industrial (C&I) customers, they can adopt automated energy management systems for buildings or factories. For a residential customer, he/she can automate the household energy management system to respond to the coupon incentives offered by the LSEs

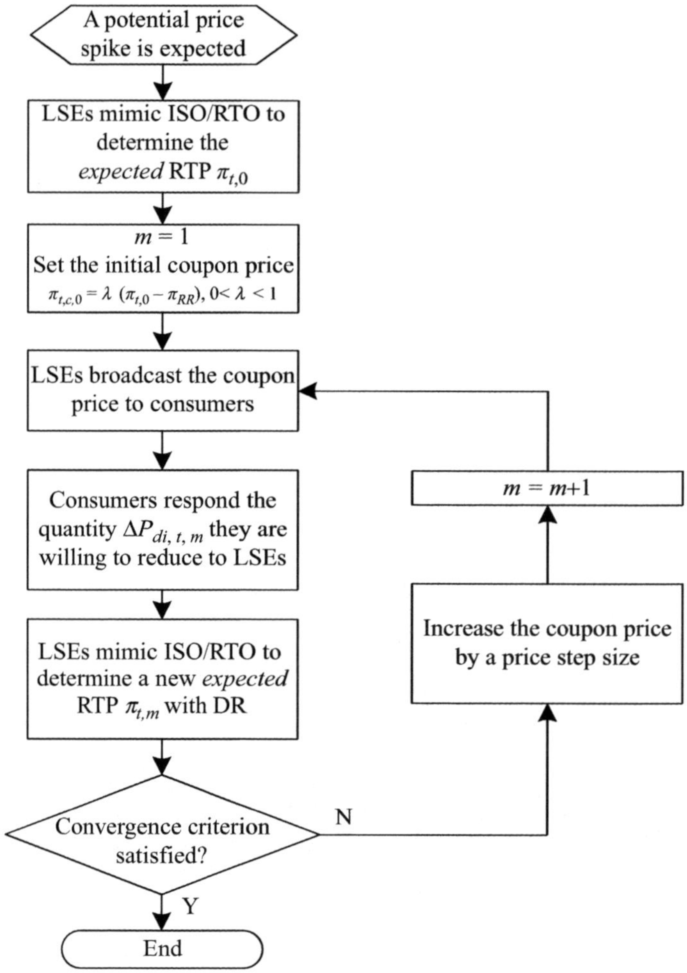

Fig. 6.3 Flowchart of coupon price adjustment for CIDR implementation

and adjust the power consumption of appliances in an automatic way. There are already some promising applications in this area, such as the SmartMeter™ of the Pacific Gas and Electric Company (PG&E) [33] and the SmartHOME™ of the Calico Energy Company [34].

6.2.8 Comparisons with the PTR/CPP (Critical Peak Pricing)

1. Comparisons with the PTR: The CIDR and the PTR have common features: (1) they are both voluntary DR programs; (2) consumers are paid to reduce their

electricity consumption; and (3) baseline calculations are needed. These common features make the critical assessment approach of this chapter applicable to the PTR program as well.
However, two fundamental features differentiate the proposed CIDR scheme from existing PTR rates.

First, the rebate rates in the PTR during critical periods are predetermined and fixed [21, 22], whereas the coupon price is determined through the iterations of LSEs and consumers based on near-real-time system conditions. Much tighter information feedback is utilized in CIDR to improve economic efficiency.

Second, the notification time for coupons is shorter than the notification time with the PTR. Typically, the PTR event is notified day-ahead. CIDR provides a mechanism to respond to a more near-term sudden change in system conditions and/or price spikes.

2. Comparisons with the CPP: First, the CPP utilizes mandatory high prices to motivate consumers to adjust their electricity consumption. This will raise equity problems. The CIDR is voluntary. Participating customers can only save money from the program.

   Second, the frequency of the CPP event is typically limited within a year [41]. For CIDR, it can be activated in a more flexible way. There is no specific limit on how often the CIDR can be activated.

   Third, the notification time in CIDR is shorter compared with that in the CPP, and the event interval is flexible in CIDR, but fixed in the CPP.

6.3 Critical Assessment of CIDR

In this section, we critically assess the social welfare, LSE profit, consumer surplus, the robustness of the retail rate, and readiness for implementation under the CIDR scheme. They are also compared with the flat rate and the RTP scheme. The analysis is based on the schematic diagram in Fig. 6.4.

The original supply function is defined as $P = f_o(\pi)$ and the inverse supply function is $\pi = f_o^{-1}(P)$. The inverse supply function is assumed to be monotonically increasing.

Suppose that 1 h ahead of an operating interval, there is a sudden wind ramp down; in this case, the updated supply function becomes $P = f_s(\pi)$. The inverse supply function with a sudden wind ramp is $\pi = f_s^{-1}(P)$. In Fig. 6.4, the original supply curve shifts to the left [42]. Given the same amount of quantity P_0, $f_s^{-1}(P_0) \geq f_o^{-1}(P_0)$.

The inverse demand function is assumed to be monotonically decreasing.

(P_{d1}, π_1) denotes the intersection between the demand curve and original supply curve. (P_{d2}, π_2) denotes the intersection between the demand curve and updated supply curve. (P_{d1}, π_3) denotes the operating point if there is no demand response. It becomes desirable to issue coupon incentives when the expected real-time wholesale price π_2 is greater than the flat retail rate π_{RR}.

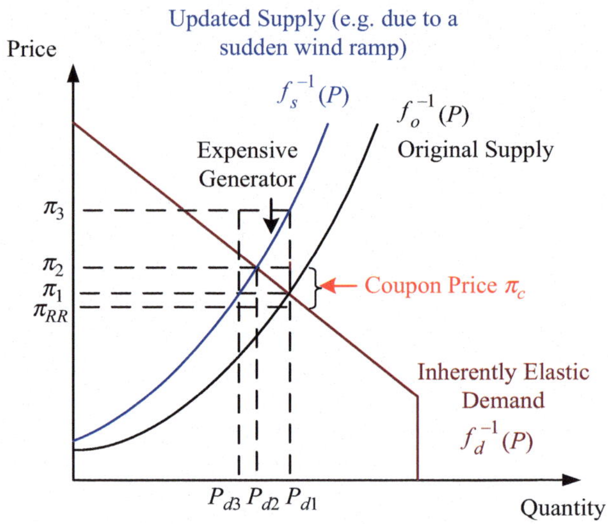

Fig. 6.4 Schematic diagram to illustrate the CIDR programs

It is assumed that under the CIDR scheme, there exists an optimal quantity P_{d3} corresponding to the optimal coupon price that maximizes LSE profit. As long as there is DR, P_{d3} is less than P_{d1}.

6.3.1 Consumer Surplus

Consumer surplus is defined as the area beneath the demand curve between 0 and the transacted quantity in Fig. 6.4. The consumer surpluses under the different schemes are listed in the following:

$$CS_{FR} = \int_0^{P_{d1}} \left(f_d^{-1}(P) - \pi_{RR}\right) dP \tag{6.11}$$

$$CS_{RTP} = \int_0^{P_{d2}} \left(f_d^{-1}(P) - f_d^{-1}(P_{d2})\right) dP \tag{6.12}$$

$$\begin{aligned} CS_{CIDR} = & \int_0^{P_{d3}} \left(f_d^{-1}(P) - \pi_{RR}\right) dP \\ & + \int_{P_{d3}}^{P_{d1}} \left(f_d^{-1}(P_{d3}) - \pi_{RR}\right) dP \end{aligned} \tag{6.13}$$

Since $f_d^{-1}(P_{d2}) = \pi_2 > \pi_{RR}$, hence $CS_{FR} > CS_{RTP}$.

Since $P_{d1} > P_{d3}$, hence $CS_{CIDR} > CS_{FR}$.

Therefore, under the CIDR scheme, consumers enjoy a higher surplus than under the flat rate and the RTP scheme.

6.3.2 LSE Profit

The different levels of LSE profit under the different schemes are listed in the following:

$$LP_{FR} = -\int_{0}^{P_{d1}} \left(f_s^{-1}(P_{d1}) - \pi_{RR}\right) dP \tag{6.14}$$

$$LP_{RTP} = 0 \tag{6.15}$$

$$\begin{aligned} LP_{CIDR} = &-\int_{0}^{P_{d3}} \left(f_s^{-1}(P_{d3}) - \pi_{RR}\right) dP \\ &-\int_{P_{d3}}^{P_{d1}} \left(f_d^{-1}(P_{d3}) - \pi_{RR}\right) dP \end{aligned} \tag{6.16}$$

Since consumers under the RTP scheme pay electricity bills at the real-time price, the financial risks are transferred from LSEs to the consumers. Without loss of generality, we assume in this formulation that LSE short-term profit is zero (the rate-of-return on investment is neglected in the short-term profit calculation).

Since $P_{d2} < P_{d1}$, hence $f_s^{-1}(P_{d2}) < f_s^{-1}(P_{d1})$ and $LP_{CIDR} > LP_{FR}$.

Unlike the flat rate structure, CIDR could reduce the loss of profit LSEs suffer due to price spikes.

In this particular case, the profit in RTP is zero whereas in the other two schemes, LSEs may lose money. But this only refers to the "significant change" periods that we are studying. During most other periods, LSEs make a profit under the flat rate.

6.3.3 Social Welfare

The social welfare is defined as the sum of the surplus of all the market participants [44]. There are three categories of profit-seeking participants: the generation companies (GenCos), the LSEs, and the end consumers.

GenCos benefit is the revenue from LSEs minus the production cost. LSE benefit is the retail revenue from consumers minus the purchasing cost from the wholesale market and the coupon payment. Consumer benefit is the total utility minus the purchasing cost from LSEs, plus the coupon rebates from LSEs.

Based on the above definitions, the social welfare is the consumers' total utility minus the GenCos' production cost.

The social welfares under the different schemes are listed in the following:

$$SW_{FR} = \int_0^{P_{d2}} (f_d^{-1}(P) - f_s^{-1}(P))dP - \int_{P_{d2}}^{P_{d1}} (f_s^{-1}(P_d) - f_d^{-1}(P))dP \tag{6.17}$$

$$SW_{RTP} = \int_0^{P_{d2}} (f_d^{-1}(P) - f_s^{-1}(P))dP \tag{6.18}$$

$$SW_{CIDR} = \int_0^{P_{d3}} (f_d^{-1}(P) - f_s^{-1}(P))dP, P_{d3} < P_{d2} \tag{6.19}$$

$$SW_{CIDR} = \int_0^{P_{d2}} (f_d^{-1}(P) - f_s^{-1}(P))dP - \int_{P_{d2}}^{P_{d3}} (f_s^{-1}(P) - f_d^{-1}(P))dP, P_{d3} > P_{d2} \tag{6.20}$$

It can be easily found that $SW_{RTP} \geq SW_{CIDR}$. If $P_{d3} = P_{d2}$, $SW_{RTP} = SW_{CIDR}$. Since $P_{d3} < P_{d1}$, hence $SW_{RTP} \geq SW_{CIDR} \geq SW_{FR}$.

The CIDR scheme achieves a higher social welfare than the flat rate scheme. It can also be shown that under the CIDR scheme, as long as P_{d3} is close to P_{d2}, the social welfare is close to that of the RTP scheme.

6.3.4 Robustness of the Retail Rate

Under the CIDR and flat rate schemes, the retail electricity rate stays constant. Therefore, the CIDR and flat rate schemes hedge retail customers from the real-time price volatilities of wholesale markets. In contrast, the RTP scheme exposes retail customers to fluctuating wholesale electricity prices.

6.3.5 Readiness for Implementation

For practical implementation, the flat rate scheme is the easiest one.

The CIDR scheme maintains the bulk structure of flat rate, while leveraging smart grid communication technologies for voluntary demand response. The CIDR targets the regions where the monitoring/information exchange infrastructure is already in place. Many of the emerging appliances allow remote control, e.g., the remote thermostat and the remote lighting control. Therefore, the additional hardware cost involved in implementing CIDR is modest. There will be moderate software costs, such as the software needed to calculate the consumers' baselines and the software for billing customers for demand reduction. Overall, CIDR is very implementable in the near future. An LSE would have the motivation to implement CIDR due to the fact that CIDR could increase the LSE's profit. With pervasive mobile communication such as smart phones, an LSE could broadcast electronic coupons to the individual customer. The customer can respond to the coupon with the help of various automation technologies. The CIDR timeline is compatible with current market operations as well. Residential customers can also participate through aggregators. The small/medium size C&I customers can participate directly for major benefits due to the relatively large base of their power consumption.

On the other hand, both utilities and regulators are risk-averse in rolling out the RTP scheme. The vast majority of customers may have a reluctance to participate in RTP due to the high price volatility [45].

6.3.6 *Pros and Cons*

Table 6.1 summarizes the qualitative comparisons of the different schemes.

As one can observe from Table 6.1, CIDR provides implementable demand response architecture, under which (1) social welfare improves compared with the flat rate scheme; (2) consumers are not exposed to fluctuating wholesale electricity prices; and (3) LSEs are better off in managing their profits as compared with the flat rate structure.

There are also some cons with the CIDR scheme.

First, the iteration procedure might increase the burden of communications and keep some potential responsive customers away from the CIDR program. The situation would be better if some automation were introduced into the procedure. As LSEs accumulate more experience, the optimal coupon pricing strategy could be estimated based on studying the collected retail customer data. Then a smaller number of iterations can be expected.

Second, more uncertainty is brought about by the CIDR scheme. Given that the CIDR scheme is a voluntary type of DR program, there will be more uncertainty compared with direct load control (DLC), interruptible load contract (ILC), and so on. However, a pilot program can be implemented to test the feasibility of the CIDR scheme. Based on the statistics, LSEs can estimate the consumers' willingness to participate. The uncertainty will be manageable if LSEs can understand the inherent behavior of customers in more detail. When CIDR becomes more reliable, it can serve as the system reserve to some extent.

Table 6.1 Qualitative comparisons of the different schemes

Hour 19 (Peak)	Flat Rate	RTP	CIDR
Consumer surplus	M	L	H
LSE profit	L	H	M
Social welfare	L	H	H
Robustness of retail rate	H	L	H
Readiness for implementation	H	L	M

Note: *H* denotes high, *M* denotes medium, *L* denotes low

Third, in terms of implementation, there will be some software costs, such as the software needed to calculate the consumers' baselines and the software for billing customers for demand reduction, etc.

6.4 Numerical Case Study

In this section, the performance of the CIDR scheme is evaluated based on realistic supply and demand data obtained from ERCOT [46].

Suppose that at 8:00 CDT, December 1, 2011, ERCOT disclosed that the wind power generation would be 1082 MW less than forecasted at 10:00 CDT. The gate closure for that 15-min settlement interval is 9:00 CDT.

The aggregated original supply curve, updated supply curve, and demand curve are illustrated in Fig. 6.5.

The shape of the curves is a staircase. No transmission congestion is assumed. Since the wind power is cheap, the supply curve above the wind power offers price shifts leftward by a quantity of 1082 MW.

All the programs are written in C language, and the optimization problems (6.1)–(6.3) and (6.9)–(6.10) are solved by CPLEX solver [47].

Three cases are considered as follows:

Case 6.1: Reference Case (Flat Rate) Consumers' inherent demand elasticity is not utilized due to the flat electricity retail rate. This case represents the status quo in which consumers do not have any incentive to respond to price signals from the market. This serves as a benchmark against which the performance of other cases is evaluated.

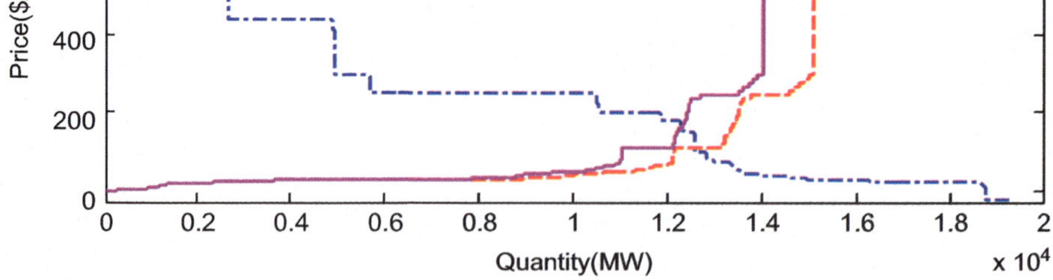

Fig. 6.5 Supply curve and demand curve of ERCOT at 10:00 CDT, on December 1, 2011

Table 6.2 Comparisons of the different schemes for one settlement interval ending at 10:15 CDT

	Case 6.1	Case 6.2	Case 6.3
	Reference case	RTP	CIDR
Peak demand(MW)	12,589	12,293	12,291
Demand response(MW)	0	295.9	298.1
Purchasing cost($)	755,350	491,730	491,642
Retail revenue($)	346,202	491,730	338,004
Coupon payment($)	0	0	4472
LSE's cost($)	755,350	491,730	496,114
Average cost($/MWh)	240	160	161
LSE's losses($)	409,148	0	158,110
Real-time Price ($/MWh)	240	160	160
Price consumers face($/MWh)	110	160	110
Consumer's rebate($)	0	0	4472
Consumer surplus($)	1,109,624	952,929	1,111,036
Social welfare($)	1,326,278	1,329,864	1,329,861

In the reference case, the retail rate of electricity is $110/MWh [48], which corresponds to in Fig. 6.4. The electricity demand equals 12,589 MW. For the sake of simplicity, the consumer baseline levels for all the cases are set at the actual demand levels of the reference case.

Case 6.2: RTP Consumers are exposed to the wholesale RTP. It is assumed that all consumers will respond to the wholesale RTP according to their respective demand bidding curves.

Case 6.3: CIDR Consumers face a flat retail rate whereas the demand reductions are paid by LSEs at coupon prices. It is assumed that consumers will respond to the coupon incentives according to their respective willingness to pay.

Table 6.2 summarizes the results for all the cases during the 15-min settlement interval ending at 10:15 CDT, on December 1, 2011.

The average cost of electricity is defined as follows:

$$\text{Average Cost} = \frac{\text{Purchasing Cost} + \text{Coupon Payment}}{\text{Peak Demand}} \tag{6.21}$$

In Case 6.1 (the reference case), all retail customers are insensitive to the wholesale real-time market price. The average cost of electricity is $240/MWh, which corresponds to in Fig. 6.4. The uniform flat retail rate would stay at $110/ MWh. Hence, the LSEs would lose $409,148 for that settlement interval.

In Case 6.2, the peak demand is reduced by 2.35% to 12,293 MW for a net social welfare increase of $3577. The average cost of electricity is $160/MWh, which corresponds to in Fig. 6.4. Since the consumers face the wholesale RTP, they are exposed to the high price and will be exposed to high price volatility. The LSEs will not lose money in this case.

In Case 6.3, the peak demand is reduced by 2.37% to 12,291 MW. The average cost of electricity is $161/MWh. This is 32.91% lower than the $240/MWh in Case 6.1, and the LSEs would decrease their losses by 61.36% for that settlement interval. The consumer surplus increases by $1412 compared to Case 6.1. Although the retail revenue decreases compared with Case 1, the LSEs' cost is reduced even more. Overall, LSE losses decrease.

In terms of peak load reduction, effective demand reduction can be achieved in Case 6.3. Hence, it will relieve the stress of power grid operation and improve system security.

In terms of social welfare, the social welfare in Case 6.3 is very close to that in Case 6.2. The social welfares of Cases 6.1 and 6.2 are shown in Figs. 6.6 and 6.7. The flat rate scheme achieves the least social welfare (Part A minus Part B in Fig. 6.6).

The CIDR scheme shows good convergence performance. From Fig. 6.8, it can be seen that with an increase in the coupon price, LSE profit under the CIDR scheme converges to the optimal point within seven iterations. With the near-real-time communication capabilities, each iteration could be done in less than 5 min. Hence, the whole process could be finished within 45 min and before the gate closure of the real-time market. This result verifies the feasibility of the timeline proposed in Sect. 6.2.

In this case, the step size of the coupon price is $10/MWh. The optimal coupon price under CIDR is $60/MWh. We will study the impact of the step size on the speed of convergence in our future work.

Compared with the fixed rebate rate of $350/MWh experimented with in Anaheim [21] and $300/MWh in Ontario [22], the CIDR scheme is much more flexible when it comes to the choice of optimal coupon price and the coupon price is much lower.

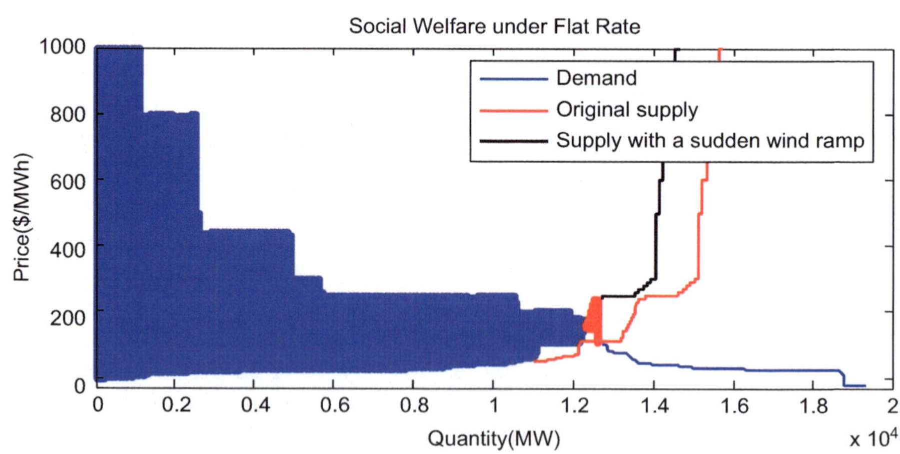

Fig. 6.6 Social welfare under the flat rate scheme

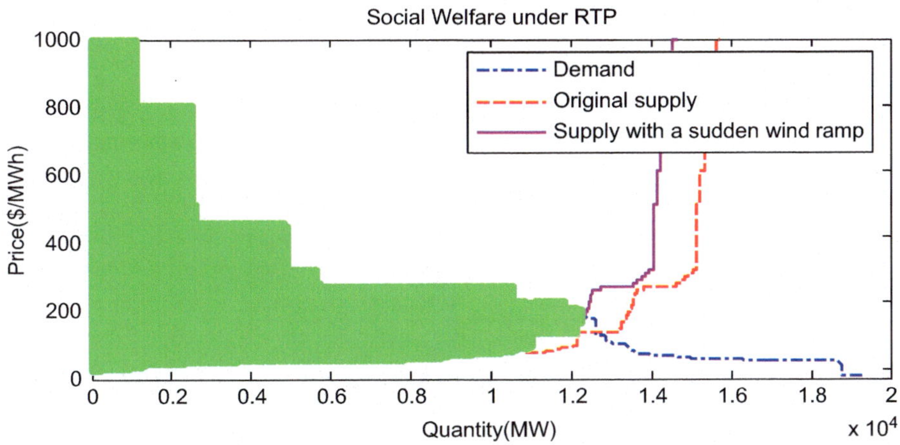

Fig. 6.7 Social welfare under the RTP scheme

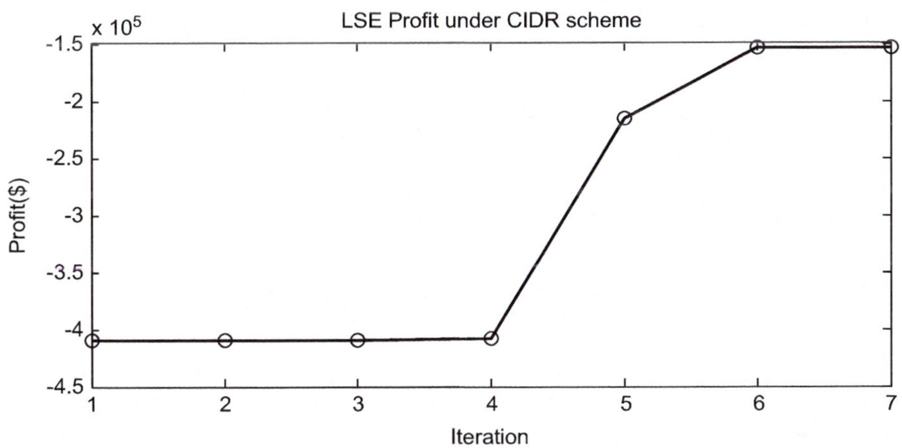

Fig. 6.8 Convergence of LSE profit under the CIDR scheme

6.4.1 Scenario Without the Iteration Between LSEs and Consumers

In practice, it may not be feasible to have multi-round iterations. Scenarios with no iteration between the LSE and consumers are tested. In this scenario, the LSE needs to estimate the coupon price ahead of time. Scenarios with different coupon prices and no iteration are tested, and the results are shown in Table 6.3.

It should be noted that even if there is no iteration between the LSE and consumers, the demand response can still bring benefits to the consumers, the LSE, and the system as a whole. The effectiveness of the CIDR scheme depends to a large extent on the LSE's ability to determine a rational coupon price based on knowledge of the consumers and system conditions.

Table 6.3 Comparisons of different coupon prices

Coupon Price($/MWh)	Demand Response(MW)	LSE's Losses($)	Coupon Payment($)	Social Welfare($)
0	0	409,148	0	1,326,287
10	2	409,082	5	1,326,354
20	4	409,028	22	1,326,417
30	7	408,984	49	1,326,474
40	206	218,767	2057	1,329,584
50	296	157,364	3699	1,329,864
60	298	158,110	4472	1,329,861

Table 6.4 Comparisons of different consumer participation rates

Participation Rate(%)	Demand Response(MW)	LSE's Losses($)	Coupon Payment($)	Social Welfare($)
100	298	158,110	4472	1,329,861
90	268	158,035	4024	1,329,794
80	238	188,837	3577	1,329,664
70	209	219,789	3130	1,329,509
60	179	219,863	2683	1,329,293
50	149	251,038	2236	1,329,040
40	119	313,537	1789	1,328,675
30	89	345,084	1341	1,328,199
20	60	408,104	894	1,327,612
10	30	408,626	447	1,326,949

6.4.2 *Impact of the Consumer Participation Rate*

In practice, it is likely that only a portion of consumers will participate. Scenarios with different consumer participation rates are tested to determine the participation rates' impact on the effectiveness of the CIDR scheme.

It can be noted from Table 6.4 that with an increase in the participation rate, the demand response will also increase, which can bring more social welfare to the whole system. A participation rate of 50% is used in the evaluation of PTR [49]. As one can observe, if the participation rate is 50% under the CIDR scheme, the LSE could decrease losses by 38.57%, from $408,626 to $251,038.

6.5 Conclusion

This chapter introduces a type of DR program targeted at retail customers who are equipped with smart meters yet still paying a flat electricity rate. This scheme is referred to as CIDR. LSEs offer coupon incentives to retail customers in anticipation of intermittent generation ramping and/or price spikes. In the CIDR scheme, the retail customers' inherent flexibility is utilized, while, at the same time, their base consumption is not exposed to wholesale real-time price fluctuations. The mechanism and algorithm designs are presented. The pros and cons are discussed in detail. Theoretical analysis and numerical case study show that CIDR can effectively induce inherent demand flexibility and reduce system-wide operational cost while maintaining a basic flat rate structure on the retail level. The CIDR scheme is compatible with today's electricity tariff structure, so it would be easily deployable on a large scale in smart grid operations of the future.

References

1. U.S. Department of energy. (2006). Benefit of demand response in electricity market and recommendations for achieving them.
2. National Action Plan for Energy Efficiency. (2010). Coordination of energy efficiency and demand response, prepared by Charles Goldman (Lawrence Berkeley National Laboratory), Michael Reid (E source), Roger Levy, and Alison Silverstein.
3. Rahimi, F., & Ipakchi, A. (2010). Demand response as a market resource under the smart grid paradigm. *IEEE Transactions on Smart Grid, 1*(1), 82–88.
4. Federal Energy Regulatory Commission. (2006). *Assessment of demand Response & Advanced Metering*. Staff Report.
5. Severin, B., Michael, J., Arthur, R., & Pricing, D. (2002). *Advanced metering and demand response in electricity markets*. Berkeley, CA: Univ. California Energy Institute.
6. Kang, C., & Jia, W. (2011). Transition of tariff structure and distribution pricing in China. In *Proceedings of the IEEE Power Engineering Society*. General Meeting.
7. (2011). *2010 California Statewide Non-Residential Critical Peak Pricing Evaluation*. Freeman Sullivan & Co.
8. Allcott, H. (2009). *Real time Pricing and electricity markets, working chapter*. Cambridge, MA: Harvard Univ.
9. Retrieved from http://www.comverge.com/newsroom/comvergepress-releases/2007/ComEd-s-Retail-Real-Time-Pricing-Program-Expa.
10. Miu-Miller, K. Integration of smart grid enabling technologies within power distribution systems. In: *Proceedings of the 2011 Japan-America Frontiers of Engineering Symposium*.
11. Sheen, J. N., Chen, C. S., & Yang, J. K. (1994). Time-of-use pricing for load management programs in Taiwan power company. *IEEE Transactions on Power Apparatus and Systems, 9* (1), 388–395.
12. Roozbehani, M., Dahleh, M. A., & Mitter, S. K. Volatility of power grids under real-time pricing. *IEEE Transactions on Power Apparatus and Systems*.
13. Cappers, P., Goldman, C., & Kathan, D. (2010). Demand response in U.S. electricity markets: Empirical evidence. *Energy, 35*(4), 1526–1535.
14. Chen, C. S., & Leu, J. T. (1990). Interruptible load control for Taiwan power company. *IEEE Transactions on Power Apparatus and Systems, 5*(2), 460–465.
15. Tan, C. W., & Varaiya, P. (1993). Interruptible electric power service contracts. *Journal of Economic Dynamics and Control, 17*, 495–517.
16. Gedra, T. W., & Varaiya, P. (1993). Markets and pricing for interruptible electric power. *IEEE Transactions on Power Apparatus and Systems, 8*(1), 122–128.
17. Tuan, L. A., & Bhattacharya, K. (2003). Competitive framework for procurement of interruptible load services. *IEEE Transactions on Power Apparatus and Systems, 18*(2), 889–897.
18. California ISO 2003 Summer Assessment. [Online]. Retrieved from http://www.caiso.com

19. Bhattacharya, K., Bollen, M. H. J., & Daalder, J. E. (2000). Real time optimal interruptible tariff mechanism incorporating utility-customer interactions. *IEEE Transactions on Power Apparatus and Systems, 15*(2), 700–706.
20. Ghosh, S., Kalagnanam, J., Katz, D., Squillante, M., Zhang, X., & E. Feinberg, "Incentive design for lowest cost aggregate energy demand reduction. In *Proceeding of the 1st IEEE International Conference on Smart Grid Communation.*
21. Wolak, F. A. (2006). *Retail customer response to real-time Pricing: The Anaheim critical peak Pricing experiment.*
22. Strapp, J., King, C., & Talbott S. (2007). *Ontario energy board smart Price pilot final report.* Prepared by IBM global business services and eMeter strategic consulting for the Ontario energy board.
23. Zhong, H., Xie, L., & Xia, Q. (2012). Coupon incentive-based demand response (CIDR) in smart grid. In *Proceedings of the IEEE Power Engineering Society.* General Meeting, 2012.
24. W. M. Warwick, A primer on electric utilities, deregulation, and restructuring of U.S. electricity markets, 2002.
25. King, C. S., & Chatterjee, S. (2003). Predicting California demand response. *Public Utilities Fortnightly, 141*(13), 27–32.
26. Historical RTM Load Zone and Hub Prices. [Online]. Retrieved from http:// www.ercot.com/mktinfo/prices/.
27. Ott, A. L. (2003). Experience with PJM market operation, system design, and implementation. *IEEE Transactions on Power Apparatus and Systems, 18*(2), 528–534.
28. ISO-NE 2010 Annual markets report. 2011. [online]. Retrieved from http://www.isone.com.
29. Report on Market Issues and Performance. CAISO Department of Market Analysis, 2010. [Online]. Retrieved from http://www.caiso.com.
30. Ng, K. H., & Sheblé, G. B. (1998). Direct load control-a profit-based load management using linear programming. *IEEE Transactions on Power Apparatus and Systems, 13*(2), 688–694.
31. Strbac, G., & Kirschen, D. (1999). Assessing the competitiveness of demandside bidding. *IEEE Transactions on Power Apparatus and Systems, 14*(1), 120–125.
32. ZigBee Alliance. (2008). *Zigbee smart energy profile specification.* Zigbee doc. 075356r15, rev. 15.
33. Retrieved from http://www.pge.com/includes/docs/pdfs/myhome/saveenergymoney/energysavingprograms/han/.
34. Retrieved from http://calicoenergy.com/.
35. Kirschen, D., Strbac, G., Cumperayot, P., & Mendes, D. (2000). Factoring the elasticity of demand in electricity prices. *IEEE Transactions on Power Apparatus and Systems, 15*(2), 612–617.
36. Chao, H.-P. (2011). Demand response in wholesale electricity markets: The choice of customer baseline. *Journal of Regulatory Economics, 39*, 68–88.
37. Nodal Protocol of ERCOT. [Online]. Retrieved from http://www.ercot.com/mktrules/nprotocols/current
38. Market Processes of California ISO. [Online]. Retrieved from http://www.caiso.com/market/Pages/MarketProcesses.aspx.
39. Gross, G., & Galiana, F. D. (1987). Short-term load forecasting. *Proceedings of the IEEE, 75* (12), 1558–1573.
40. Retrieved from http://www.sce.com/b-rs/demand-response-programs/demand-response-programs.htm.
41. The demand response baseline. White chapter, EnerNOC Inc., 2008.
42. Kirschen, D. S. (2003). Demand-side view of electricity markets. *IEEE Transactions on Power Apparatus and Systems, 18*(2), 520–527.
43. Zimmerman, R. D., Murillo-Sânchez, C. E., & Thomas, R. J. (2011). MATPOWER: Steady-state operations, planning, and analysis tools for power systems research and education. *IEEE Transactions on Power Apparatus and Systems, 26*(1), 12–19.

44. Schweppe, F. C., Caramanis, M., Tabors, R., & Bohn, R. (1988). *Spot pricing of electricity*. Boston, MA: Kluwer.
45. Faruqui, A., & Mauldin, M. (2002). The barriers to real-time pricing: Separating fact from fiction. *Public Utilities Fortnightly, 140*(14), 30–40.
46. Retrieved from http://mis.ercot.com/misapp/GetReports.do?reportTypeId=13054&reportTitle=48 Hour DAM and SCED Energy Curves Reports&showHTMLView=&mimicKey.
47. The IBM ILOG CPLEX Website, 2011. [Online]. Retrieved from http://www-01.ibm.com/software/websphere/products/optimization/academicinitiative/index.html/.
48. Annual Report of ERCOT 2009. [Online]. Retrieved from http://www.ercot.com.
49. Ex ante load impact forecast for SDG&E Peak Time Rebate. 2010.

Chapter 7
Distributed Real-Time Demand Response

7.1 Introduction

7.1.1 Motivation

With the rapid growth of smart grid technology, an increasing number of distributed energy resources (DERs) will be integrated into distribution grids, including distributed generation (DG), microgrids (MGs), energy storage systems (ESSs), electric vehicles (EVs), and demand response (DR) resources [1, 2]. It is a difficult task to coordinate the ubiquitous distributed energy resources in a centralized manner, mainly due to the heavy communication burden and computational costs. Moreover, the details of consumers' privacy are required in the centralized optimization algorithm, which is scarcely achievable under *information asymmetry*.

By contrast, distributed optimization algorithms merely need a small amount of consumers' information, keeping the consumers' privacy. In addition, distributed algorithms can decompose a large-scale model into a series of subproblems. These more tractable subproblems are then solved iteratively in an independent and parallel manner, which greatly improves the overall efficiency. Therefore, distributed decision making has attracted growing interests around the entire world.

However, the convergence performance of existing distributed algorithms highly relies on the selection of the iteration step size and search direction, such as the subgradient projection method [3–5]. Without fully using iterative information, oscillations may occur when coordinating the master problem and subproblems, which dramatically lowers the efficiency. To this end, the Lagrangian multiplier optimal selection (LMOS) approach is presented in this chapter to overcome the difficulty of choosing the iteration step size or search direction. By estimating the energy demand elasticity of consumers, the Lagrangian multipliers are optimized during each iteration, which greatly accelerates the convergence.

P. Du et al., *Demand Response in Smart Grids*,
https://doi.org/10.1007/978-3-030-19769-8_7

7.1.2 Literature Review and Contribution

Demand response has become one of the most important measures to address the ever increasing peak load [6–8]. In DR programs, consumers are inclined to reduce their electricity consumption during critical peak periods or shift some of the peak demand to off-peak periods [9, 10]. Thereby, the economy and security of power systems are improved [11]. In recent decades, real-time DR and distributed algorithms have been widely investigated [12–14]. In [15], DR has been shown to be a promising tool for balancing generation and demand and facilitating renewable energy accommodation. The potential for the implementation of price-based DR by industrial consumers to increase their use of wind power is analyzed. In [16], the optimal installed capacity allocation of renewable resources is investigated in conjunction with DR resources. Then an integrated model is established to determine a cost-minimizing allocation of renewable asset investments. In [17], a distributed real-time DR algorithm is proposed to determine the demand of multiple users and the supply from multiple utilities. By decoupling the original model into single-seller-multi-buyer subsystems, each subsystem can be solved in a distributed manner. In [18], a distributed direct load control scheme for large-scale residential DR is proposed. An average consensus algorithm is utilized to optimize the consumption of each user and schedule in-building appliance use. In [19], a coordinated DR scheme is formulated to mitigate severe peak rebounds in low price periods. An iterative distributed algorithm is proposed to manage the DR provided by residential customers. In [20], a min-max-min cost model is proposed to find a robust optimal day-ahead scheduling of smart distribution networks. In the model, DR programs are taken as important resources, participating in both energy and reserve scheduling.

The selection of coordinated Lagrangian multipliers has a great influence on the convergence. Thus, extensive research has been focused on the improvement of the Lagrangian multiplier. In [21], an enhanced adaptive Lagrangian relaxation (ELR) for unit commitment is presented. The Lagrangian multipliers are updated adaptively and heuristically, leading to much fewer iterations until convergence than when using the sub-gradient method. In [3], the dynamic economic dispatch model is decoupled into a two-stage dual problem via Lagrangian relaxation. The Lagrangian multipliers are then updated based on a quasi-Newton method. In [4], a dynamic multiplier-based method is presented. In the method, Lagrangian multipliers are derived as the linear functions of the tie-line flow deviation. However, the linear functions cannot provide the optimal projection directions. In [5], a binary search-based approach is proposed to accelerate the convergence of the distributed algorithm. However, congestion multipliers still rely on the sub-gradient projection method. According to the literature review, if the projection vectors are chosen improperly, the number of iterations will increase dramatically. Sometimes the convergence of the algorithm cannot even be guaranteed. Therefore, how to determine optimal Lagrangian multipliers still needs an in-depth study.

To the best of our knowledge, the Lagrangian multiplier optimal selection approach is proposed for the first time in this chapter. By applying Lagrangian relaxation (LR), the real-time DR model is decoupled into several single-consumer models. According to the RTPs and price prediction, each consumer optimizes the operation of appliances and sends back the scheduled load profiles to the load serving entity (LSE). The Lagrangian multipliers are then determined by the LSE in an optimization manner during each iteration. Via sensitivity analysis, the LSE can effectively estimate the energy demand elasticity of the consumers. Based on the demand elasticity, the LSE optimizes the LMOS model and broadcasts the Lagrangian multipliers. The LMOS approach provides the LSE with precise price signals to leverage the balance between energy supply and demand.

The major contributions of the work described in this chapter are as follows:

1. The framework of distributed real-time DR based on LMOS approach is constructed. In this framework, the real-time DR model is decoupled and optimized in a distributed and iterative manner via information exchange between the LSE and consumers. In each iteration, consumers schedule their energy demands for the next few hours and send their energy demands to the LSE. Then, the LSE optimizes Lagrangian multipliers for the next iteration and broadcasts the Lagrangian multipliers to each consumer.
2. The LMOS approach is proposed and applied for accelerating the convergence of the distributed algorithm. By utilizing the proposed approach, the difficulty of choosing the search direction and the step size is overcome. According to the theoretical derivation of sensitivity analysis, energy demand elasticity of consumers can be effectively estimated, and the LMOS model is then established. By optimizing this model, the LSE is able to determine optimal Lagrangian multipliers in a linearized neighborhood, which dramatically improves the convergence.

The rest of this chapter is organized as follows: In Sect. 7.2, the mathematical model of real-time DR is presented. Then the DR model is decoupled into several single-consumer subproblems using the Lagrangian relaxation approach in Sect. 7.3. In Sect. 7.4, sensitivity analysis of a general convex optimization problem is studied. In Sect. 7.5, based on the objective and variable sensitivity, the LMOS model is established to optimally determine the Lagrangian multipliers during iterations. In Sect. 7.6, case studies based on a distribution grid with the number of consumers ranging from 10 to 100 are utilized to demonstrate the efficiency and robustness of the proposed approach. Section 7.7 concludes the study.

7.2 Real-Time Demand Response Model

In a smart distribution grid, the advanced metering infrastructure (AMI) makes bidirectional communication possible between load-serving entities (LSEs) and various types of consumers. The contractual arrangement allows consumers to receive real-time price (RTP) signals several minutes prior to a given hour and respond to the signal by subsequently adjusting their electricity consumption [22]. A

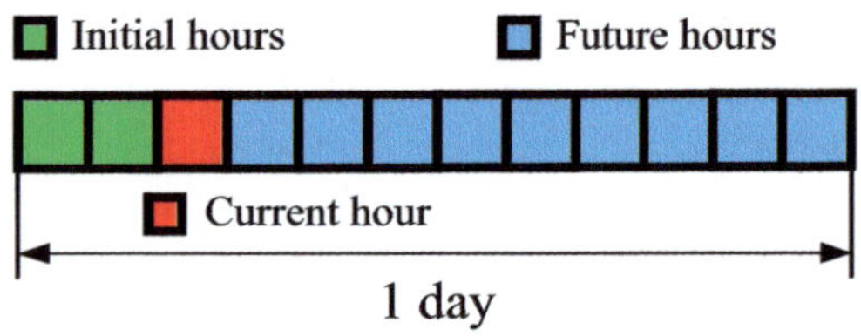

Fig. 7.1 The schematic of real-time DR

rational consumer is supposed to maximize his/her utility by optimizing his/her electricity consumption profile. Because real-time pricing reflects the time-varying marginal costs of generating and delivering electric power, RTP is an effective measure to achieve the economical operation of power grids.

The availability of RTP makes consumers aware of hourly price information several minutes prior to the current hour T_S. Then, consumers can optimize their energy demands for the following hours in a rolling manner. It is assumed that (1) prices and energy consumption for the initial $T_S - 1$ h are determined and known by each consumer; (2) the prices for the current hour T_S and the following hours are forecasted and broadcast by the LSE; and (3) consumers optimize the demand profile for the next few hours. The schematic of real-time DR is shown in Fig. 7.1. Given the energy consumption for the initial $T_S - 1$ h, the consumer will optimize the demands for the current hour T_S and the following hours based on forecasted energy prices. Then the consumer will implement the energy consumption at T_S. At next time slot, the cost minimization process will continue in a rolling manner according to the updated initial energy consumption and forecasted prices.

It is worth mentioning that this chapter is focused on the effectiveness and efficiency of the proposed distributed solution framework. Thus, the uncertainty of price forecasting [23–25] is ignored; real-time distributed DR considering price uncertainty is worthy of further research.

7.2.1 *Objective*

Consider a smart distribution grid with an LSE and a number of consumers from 1 to N. A day is composed of a set of discrete time slots from 1 to T. In this chapter, the LSE serves as an agent to maximize the consumers' surplus. Consumers optimize their energy demand for the current hour T_S and the following $T - T_S$ hours by referring to utility functions that quantify the comfort of each consumer. As a basic assumption in microeconomics, the utility functions of the consumers are assumed to be quadratic and concave [17].

Because the energy consumption of the initial $T_S - 1$ h is determined, the surplus of the initial $T_S - 1$ h is neglected. Therefore, the objective is to maximize the total surplus of all consumers from the current hour until the end of the day,

$$\max \quad W(\boldsymbol{P}) = \sum_{i=1}^{N} \sum_{t=T_S}^{T} W_i^t(p_i^t) \tag{7.1}$$

where $W(\cdot)$ is the surplus function of all consumers, $\boldsymbol{P}$ is energy demand matrix of the consumers, p_i^t is the continuous variable of the energy demand of consumer i in time slot t, and $W_i^t(\cdot)$ is the surplus function of consumer i in time slot t, i.e., the utility function minus the electricity purchase costs,

$$W_i^t(p_i^t) = U_i^t(p_i^t) - C^t p_i^t \tag{7.2}$$

where $U_i^t(\cdot)$ is the utility function [5] of consumer i in time slot t, and C^t is the RTP in time slot t. The utility is a measure of preference over some goods and services, which represents the satisfaction experienced by consumers [26]. In practice, utility functions can be obtained by quantifying electricity benefits, which has been widely investigated in existing literature [27]. In [28], utility data are used to estimate consumers' cost functions. The coefficients of cost functions are calibrated using real data obtained from utility demand management programs. In [29], the parameters in consumers' utility functions are acquired using maximum likelihood estimation. In [30], the consumers' utility functions are regressed using a polynomial estimation based on the interaction data between the agent and consumers. Therefore, based on the historical data, the utility functions of consumers can be effectively regressed and the corresponding parameters can be obtained.

In this chapter, it is assumed that each consumer knows the individual preferences and determines his/her utility function [27]. To preserve the privacy of consumers, the detailed utility functions will not be collected by the LSE, while LSE is able to estimate consumers' partial information for accelerating convergence.

7.2.2 *Constraints*

The real-time DR model is subject to spatial and temporal constraints.

1. The constraint of the supply capacity:

$$\sum_{i=1}^{N} p_i^t \le TP^t, \quad \forall t \tag{7.3}$$

   where TP^t is the supply capacity in time slot t. The supply capacity comes from the distribution grid limits, such as the thermal limit of the transformers.
2. The constraints of the lower and upper energy bounds:

$$\underline{p}_i^t \leq p_i^t \leq \bar{p}_i^t, \ \forall i, \ \forall t \tag{7.4}$$

where $\underline{p}_i^t$ and $\bar{p}_i^t$ are the lower and upper energy bounds of consumer i in time slot t, respectively, $\underline{p}_i^t$ is the minimal energy demand of consumer i, comprising the demand of must-run appliances, and $\bar{p}_i^t$ is generally the maximal energy consumption of all appliances of consumer i.

3. The constraints of the daily demand:

$$\sum_{t=1}^{T} p_i^t \geq TE_i, \ \forall i \tag{7.5}$$

where TE_i is the daily required energy demand of consumer i. For each consumer, the energy consumption is to be scheduled from T_S to T. The energy consumption during the initial $T_S - 1$ h is constant. These constraints guarantee the daily operation of the must-run appliances.

Therefore, the objective function (7.1) and the constraints (7.3)–(7.5) form the real-time DR model. The DR model can be easily expanded by adding other practical constraints.

7.3 Lagrangian Relaxation Approach

In a smart distribution with ubiquitous DERs, it is difficult to schedule the large amounts of DR resources in a centralized manner, mainly due to the heavy communication burden and computational costs. In addition, preserving the privacy of consumers is also an important issue. Therefore, in this chapter, the DR model is optimized in a distributed manner.

To handle spatially and temporally coupled constraints, the LR approach is adopted. The DR model is decoupled into several single-consumer subproblems. Each subproblem can be solved in an independent and parallel way.

The Lagrangian function of the DR model is defined as

$$\begin{aligned} L(\boldsymbol{P}, \boldsymbol{\Lambda}) &= \sum_{i=1}^{N} \sum_{t=T_S}^{T} W_i^t(p_i^t) - \sum_{t=T_S}^{T} \lambda^t \left(\sum_{i=1}^{N} p_i^t - TP^t \right) \\ &= \sum_{i=1}^{N} \sum_{t=T_S}^{T} \left(W_i^t(p_i^t) - \lambda^t p_i^t \right) + \sum_{t=T_S}^{T} \lambda^t \cdot TP^t, \quad \boldsymbol{P} \in \boldsymbol{X} \end{aligned} \tag{7.6}$$

where $L(\boldsymbol{P}, \boldsymbol{\Lambda})$ is the Lagrangian function, $\boldsymbol{\Lambda}$ is the Lagrangian multiplier vector of constraint (7.3), i.e., $\boldsymbol{\Lambda} = \left[\lambda^{T_S}, \ldots \lambda^{T}\right]^{\mathrm{T}}$, and $\boldsymbol{X}$ is the constraint set of $\boldsymbol{P}$, including all the individual constraints of consumers.

The dual function is the objective of the dual problem, i.e., the supreme of the Lagrangian function over the energy demand $\boldsymbol{P}$. Because each consumer makes his/her own decision, the dual function can be decoupled into the sum of all consumers' objectives,

$$\begin{aligned} D(\boldsymbol{\Lambda}) &= \sup_{\boldsymbol{P}\in \boldsymbol{X}} \ L(\boldsymbol{P},\boldsymbol{\Lambda}) \\ &= \sum_{i=1}^{N}\sum_{t=T_S}^{T} S_i^t(\lambda^t) + \sum_{t=T_S}^{T} \lambda^t \cdot TP^t \end{aligned} \tag{7.7}$$

where $D(\Lambda)$ is the dual function and $S_i^t(\lambda^t)$ is the maximization of consumer i's surplus in time slot t,

$$\begin{aligned} S_i^t(\lambda^t) &= \max_{p_i^t \in \boldsymbol{X}_i} \ B_i^t(p_i^t, \lambda^t) \\ &= \max_{p_i^t \in \boldsymbol{X}_i} \ W_i^t(p_i^t) - \lambda^t p_i^t \end{aligned} \tag{7.8}$$

where X_i is the constraint set of consumer i, and $B_i^t(p_i^t, \lambda^t)$ is expressed as $W_i^t(p_i^t) - \lambda^t p_i^t$. Because the Lagrangian multiplier λ^t reflects the usage of the supply capacity, λ^t is defined as the congestion price in the existing literature [31]. The physical meaning of eq. (7.8) is that each consumer maximizes the surplus during each time slot considering the congestion prices. It is reasonable to assume that the consumers would like to interact with the LSE and maximize his/her surplus taking Lagrangian multipliers as congestion prices, which is common in existing researches [5, 12, 17].

Therefore, the dual problem is to minimize the dual function over the Lagrangian multiplier $\boldsymbol{\Lambda}$.

$$\min_{\Lambda \geq 0} \ D(\boldsymbol{\Lambda}) \tag{7.9}$$

According to the *strong duality principle*, the primal and dual problem are strictly equivalent. The dual problem can be solved iteratively by the distributed algorithm proposed in this chapter. From the optimal solution $\boldsymbol{\Lambda}^*$, the optimal solution to consumers' energy demands $\boldsymbol{P}^*$ can be obtained by solving the problem (7.8).

7.4 Sensitivity Analysis

The effectiveness and efficiency of existing distributed algorithms greatly rely on the choice of the iteration step size and search direction. Without making full use of iterative information, oscillations may occur when coordinating the LSE and consumers, lacking robustness and optimality. While extensive research has been focused on the improvement of Lagrangian multipliers, how to determine the optimal Lagrangian multipliers still needs an in-depth study.

To optimize Lagrangian multipliers in each iteration, sensitivity analysis can be utilized to search for the extrema of Lagrangian multipliers in a linearized space. Given the Lagrangian multipliers, the linearized neighborhood can be constructed by sensitivity analysis. The Lagrangian multipliers for next iteration can then be determined in the linearized space.

As is shown in Sect. 7.3, the subproblem of each consumer is an optimization model treating the consumers' energy demands $\boldsymbol{P}$ as optimization variables and the Lagrangian multiplier $\boldsymbol{\Lambda}$ as parametric variables. This model can be abstracted as a general convex optimization problem. In this section, the sensitivity analysis for a general convex optimization problem is derived.

7.4.1 General Sensitivity Expressions

A general convex optimization problem has the form

$$\min_{\boldsymbol{x}} \ z = f(\boldsymbol{x}, \boldsymbol{a}) \tag{7.10}$$

subject to

$$\boldsymbol{h}(\boldsymbol{x}) = 0 \tag{7.11}$$

$$\boldsymbol{g}(\boldsymbol{x}) \leq 0 \tag{7.12}$$

where $z = f(\boldsymbol{x}, \boldsymbol{a})$ is the objective function of the optimization problem, $\boldsymbol{x}$ is the variable vector, $\boldsymbol{a}$ is the parametric vector, $\boldsymbol{h}(\boldsymbol{x}) = [h_1(\boldsymbol{x}), \ldots, h_\beta(\boldsymbol{x})]^{\mathrm{T}}$ are the equality constraints, $\boldsymbol{g}(\boldsymbol{x}) = [g_1(\boldsymbol{x}), \ldots, g_\gamma(\boldsymbol{x})]^{\mathrm{T}}$ are the inequality constraints, and β and γ are the numbers of equality and inequality constraints.

The Karush-Kuhn-Tucker (KKT) first-order optimality conditions for this problem are

$$\begin{aligned} & \nabla_x f + \nabla_x \boldsymbol{h} \cdot \boldsymbol{v}^* + \nabla_x \boldsymbol{g} \cdot \boldsymbol{w}^* \\ & = \nabla_x f + \sum_{j=1}^{\beta} v_j^* \nabla_x h_j + \sum_{j=1}^{\gamma} w_j^* \nabla_x g_j \\ & = 0 \end{aligned} \tag{7.13}$$

$$\boldsymbol{h}(\boldsymbol{x}^*) = 0 \tag{7.14}$$

$$\boldsymbol{g}(\boldsymbol{x}^*) \leq 0 \tag{7.15}$$

$$\boldsymbol{w}^{*\mathrm{T}} \cdot \boldsymbol{g}(\boldsymbol{x}^*) = 0 \tag{7.16}$$

$$\boldsymbol{w}^* \geq 0 \tag{7.17}$$

where ∇ is the Laplace operator, $\boldsymbol{x}^*$ is the optimal variable vector, and $\boldsymbol{v}^*$ and $\boldsymbol{w}^*$ are the optimal Lagrangian multiplier vectors of the equality and inequality constraints, respectively.

To analyze the sensitivity when the parametric vector $\boldsymbol{a}$ is perturbed, the derivatives of eqs. (7.13)–(7.15) are

$$\begin{aligned}&\left[\nabla_{xx}f + \sum_{j=1}^{\beta} v_j^* \nabla_{xx}h_j + \sum_{j=1}^{\gamma'} w_j^* \nabla_{xx}g_j\right] \cdot d\boldsymbol{x}\\&+\nabla_{xa}f \cdot d\boldsymbol{a} + \nabla_x\boldsymbol{h} \cdot d\boldsymbol{v} + \nabla_x\boldsymbol{g}' \cdot d\boldsymbol{w}' = 0\end{aligned} \tag{7.18}$$

$$\nabla_x\boldsymbol{h}^{\mathrm{T}} \cdot d\boldsymbol{x} = 0 \tag{7.19}$$

$$\nabla_x\boldsymbol{g}'^{\mathrm{T}} \cdot d\boldsymbol{x} = 0 \tag{7.20}$$

where γ' is the number of binding (active) inequality constraints, and $\boldsymbol{g}'(\boldsymbol{x}^*)$ are the binding inequality constraints, i.e.,

$$\boldsymbol{g}'(\boldsymbol{x}^*) = 0 \tag{7.21}$$

According to eqs. (7.18)–(7.20), the matrix form is

$$\begin{bmatrix} \boldsymbol{F}_{xx} & \boldsymbol{H}_x & \boldsymbol{G}_x \\ \boldsymbol{H}_x^T & 0 & 0 \\ \boldsymbol{G}_x^T & 0 & 0 \end{bmatrix} \begin{bmatrix} d\boldsymbol{x} \\ d\boldsymbol{v} \\ d\boldsymbol{w}' \end{bmatrix} = -\begin{bmatrix} \boldsymbol{F}_{xa} \\ 0 \\ 0 \end{bmatrix} d\boldsymbol{a} \tag{7.22}$$

$$\boldsymbol{F}_{xx} = \nabla_{xx}f + \sum_{j=1}^{\beta} v_j^* \nabla_{xx}h_j + \sum_{j=1}^{\gamma'} w_j^* \nabla_{xx}g_j \tag{7.23}$$

$$\boldsymbol{F}_{xa} = \nabla_{xa}f \tag{7.24}$$

$$\boldsymbol{H}_x = \nabla_x\boldsymbol{h} \tag{7.25}$$

$$\boldsymbol{G}_x = \nabla_x\boldsymbol{g}' \tag{7.26}$$

Thus, the general sensitivity expressions are obtained. Next, the sensitivities of the objective and variables are derived.

7.4.2 Objective Sensitivity

According to eq. (7.13), the expression below holds:

$$\begin{aligned} 0 &= (\nabla_x f + \nabla_x \boldsymbol{h} \cdot \boldsymbol{v}^* + \nabla_x \boldsymbol{g} \cdot \boldsymbol{w}^*)^{\mathrm{T}} d\boldsymbol{x} \\ &= \nabla_x f^{\mathrm{T}} \cdot d\boldsymbol{x} + \boldsymbol{v}^{*\mathrm{T}} \cdot \nabla_x \boldsymbol{h}^{\mathrm{T}} \cdot d\boldsymbol{x} + \boldsymbol{w}^{*\mathrm{T}} \cdot \nabla_x \boldsymbol{g}^{\mathrm{T}} \cdot d\boldsymbol{x} \end{aligned} \tag{7.27}$$

According to differentiations of eqs. (7.19) and (7.20), eq. (7.27) can be simplified as

$$\begin{aligned} 0 &= \nabla_x f^{\mathrm{T}} \cdot d\boldsymbol{x} + \boldsymbol{v}^{*\mathrm{T}} \cdot 0 + [\boldsymbol{w}'^{\mathrm{T}} \;\; \boldsymbol{w}''^{\mathrm{T}}] \begin{bmatrix} \nabla_x \boldsymbol{g}'^{\mathrm{T}} \cdot d\boldsymbol{x} \\ \nabla_x \boldsymbol{g}''^{\mathrm{T}} \cdot d\boldsymbol{x} \end{bmatrix} \\ &= \nabla_x f^{\mathrm{T}} \cdot d\boldsymbol{x} + [\boldsymbol{w}'^{\mathrm{T}} \;\; 0] \begin{bmatrix} 0 \\ \nabla_x \boldsymbol{g}''^{\mathrm{T}} \cdot d\boldsymbol{x} \end{bmatrix} \\ &= \nabla_x f^{\mathrm{T}} \cdot d\boldsymbol{x} \end{aligned} \tag{7.28}$$

where $\boldsymbol{g}''(\boldsymbol{x}^*)$ are the inactive inequality constraints, and $\boldsymbol{w}''$ is the corresponding Lagrangian multiplier vector, equal to a zero vector because of the *complementary slackness*.

In view of eq. (7.28), the objective sensitivity equation can be obtained by differentiating eq. (7.10).

$$dz = \nabla_x f^{\mathrm{T}} \cdot d\boldsymbol{x} + \nabla_a f^{\mathrm{T}} \cdot d\boldsymbol{a} = \nabla_a f^{\mathrm{T}} \cdot d\boldsymbol{a} \tag{7.29}$$

Therefore, the objective sensitivity demonstrates how much the objective will change with the increment of parametric variables.

7.4.3 Variable Sensitivity

For simplification, $d\boldsymbol{\mu}$ is substituted for $d\boldsymbol{v}$ and $d\boldsymbol{w}'$, i.e.,

$$d\boldsymbol{\mu} = \begin{bmatrix} d\boldsymbol{v} \\ d\boldsymbol{w}' \end{bmatrix} \tag{7.30}$$

Then, eq. (7.22) can be rewritten as

$$\begin{bmatrix} \boldsymbol{F}_{xx} & \boldsymbol{E}_x \\ \boldsymbol{E}_x^{\mathrm{T}} & 0 \end{bmatrix} \begin{bmatrix} d\boldsymbol{x} \\ d\boldsymbol{\mu} \end{bmatrix} = - \begin{bmatrix} \boldsymbol{F}_{xa} \\ 0 \end{bmatrix} d\boldsymbol{a} \tag{7.31}$$

$$\boldsymbol{E}_x = [\boldsymbol{H}_x \;\; \boldsymbol{G}_x] \tag{7.32}$$

According to the inversion rule of block matrices, eq. (7.31) can be expressed as

$$\begin{aligned}\begin{bmatrix} dx \\ d\mu \end{bmatrix} &= -\begin{bmatrix} F_{xx} & E_x \\ E_x^{\mathrm{T}} & 0 \end{bmatrix}^{-1} \begin{bmatrix} F_{xa} \\ 0 \end{bmatrix} da \\ &= -\begin{bmatrix} A_{11} & A_{12} \\ A_{21} & A_{22} \end{bmatrix} \begin{bmatrix} F_{xa} \\ 0 \end{bmatrix} da \\ &= \begin{bmatrix} -A_{11}F_{xa} \\ -A_{21}F_{xa} \end{bmatrix} da \end{aligned} \tag{7.33}$$

$$A_{11} = F_{xx}^{-1} + F_{xx}^{-1}E_x\left(-E_x^{\mathrm{T}}F_{xx}^{-1}E_x\right)^{-1}E_x^{\mathrm{T}}F_{xx}^{-1} \tag{7.34}$$

$$A_{12} = -F_{xx}^{-1}E_x\left(-E_x^{\mathrm{T}}F_{xx}^{-1}E_x\right)^{-1} \tag{7.35}$$

$$A_{21} = -\left(-E_x^{\mathrm{T}}F_{xx}^{-1}E_x\right)^{-1}E_x^{\mathrm{T}}F_{xx}^{-1} \tag{7.36}$$

$$A_{22} = \left(-E_x^{\mathrm{T}}F_{xx}^{-1}E_x\right)^{-1} \tag{7.37}$$

Therefore, the variable sensitivity equation is

$$\begin{aligned} dx &= -A_{11}F_{xa}da \\ &= \left[F_{xx}^{-1}E_x\left(E_x^{\mathrm{T}}F_{xx}^{-1}E_x\right)^{-1}E_x^{\mathrm{T}} - I\right]F_{xx}^{-1}F_{xa}da \end{aligned} \tag{7.38}$$

The variable sensitivity demonstrates how much the optimum of the optimization variables will change with the increment of parametric variables.

The objective sensitivity and variable sensitivity equations are the basis of the LMOS approach. In the single-consumer subproblem, the objective sensitivity indicates the relation between the surplus of all consumers and the congestion prices, while the variable sensitivity shows the relation between the energy demands of the consumers and the congestion prices.

7.5 Lagrangian Multiplier Optimal Selection Approach

In contrast to existing distributed algorithms, the Lagrangian multipliers can be optimized during iterations using the proposed LMOS approach. From an economic perspective, the Lagrangian multipliers of the supply capacity constraints are the congestion prices in time slot t [31]. The congestion prices are determined and broadcast by the LSE to leverage the balance between supply and demand. If the energy demands of all consumers exceed the supply capacity, the LSE can raise the congestion prices to incentivize consumers to reduce their energy demands.

To get around the difficulty of selecting the iteration step size and search direction, congestion prices are optimized in a linearized space using iterative information. Given the energy demands of consumers $\boldsymbol{P}(k)$ and congestion prices $\boldsymbol{\Lambda}(k)$ in the kth iteration, the LSE needs to determine congestion prices $\boldsymbol{\Lambda}(k+1)$ in the $(k+1)$ iteration.

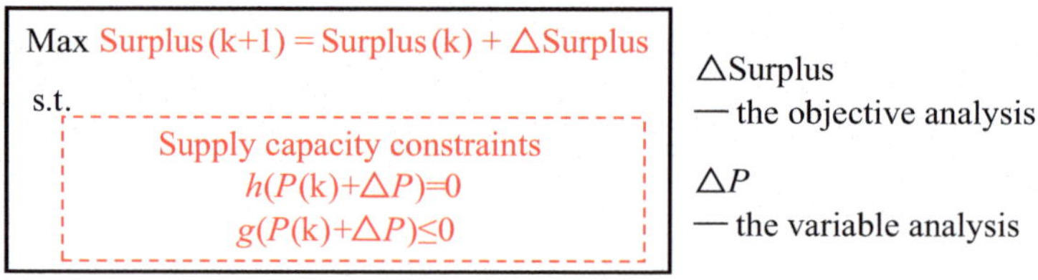

Fig. 7.2 The illustration for the LMOS model

The illustration for the LMOS model is shown in Fig. 7.2.

The objective of the LSE is to maximize the total surplus of all consumers subject to the supply capacity constraints. However, with the absence of global information, only local information can be used. Thus, sensitivity equations are applied to establish the LMOS model. For a single-consumer subproblem, the congestion price λ^t is the parametric variable. According to eq. (7.29), the objective sensitivity is

$$dB_i^t = \frac{d\left(W_i^t(p_i^t) - \lambda^t p_i^t\right)}{d\lambda^t} d\lambda^t = -p_i^t d\lambda^t \tag{7.39}$$

According to eq. (7.38) and the Appendix, the variable sensitivity is

$$dp_i^t = -\left[\frac{d^2 B_i^t(p_i^t)}{d(p_i^t)^2}\right]^{-1} \left[\frac{d^2 B_i^t(p_i^t)}{d\lambda^t dp_i^t}\right] d\lambda^t = \left[\frac{d^2 W_i^t(p_i^t)}{d(p_i^t)^2}\right]^{-1} d\lambda^t \tag{7.40}$$

Therefore, the detailed LMOS model is formulated as

$$\begin{aligned} \max_{\Lambda(k+1)} & \sum_{i=1}^{N} \sum_{t=T_S}^{T} \left(B_i^t + dB_i^t\right) \\ & = B(k) + \sum_{i=1}^{N} \sum_{t=T_S}^{T} -p_i^t(k) \cdot \left[\lambda^t(k+1) - \lambda^t(k)\right] \end{aligned} \tag{7.41}$$

subject to

$$\sum_{i=1}^{N} \left\{ p_i^t(k) + \left[\frac{d^2 W_i^t}{d(p_i^t)^2}\right]^{-1} \left[\lambda^t(k+1) - \lambda^t(k)\right] \right\} \leq TP^t, \ \forall t \tag{7.42}$$

$$\left|\lambda^t(k+1) - \lambda^t(k)\right| \leq \varepsilon, \ \forall t \tag{7.43}$$

where $B(k)$ is the objective value in the kth iteration, which is a constant and can be neglected in the $(k + 1)^{\text{st}}$ iteration, $p_i^t(k)$ is the energy demand of consumer i for time slot t in the kth iteration, and $\lambda^t(k)$ is the congestion price in time slot t in the kth

iteration. Given $p_i^t(k)$ and $\lambda^t(k)$, $\lambda^t(k+1)$ is the decision variable of the LMOS model. The objective indicates that the LSE maximizes the surplus of all consumers while optimizing congestion prices in the $(k+1)^{\text{st}}$ iteration.

The supply capacity limit in the LMOS is shown as the constraint (7.42). Because the sensitivity of Lagrangian multipliers is applied to establish the optimization model, to ensure the effectiveness of the linearized space, $d\lambda^t$ should be limited within a small range ε, as constraint (7.43) shows.

It is important to note that the second derivatives of the utility functions of each consumer are needed in the LMOS model. In practice, the second derivatives of the utility functions can be estimated based on the interactions between the LSE and consumers. As shown in eq. (7.40), the LSE is able to obtain the second derivatives of consumers by observing the energy demand increments in response to congestion price perturbations.

The framework of real-time DR based on LMOS is illustrated in Fig. 7.3.

During the kth iteration, consumers optimize their energy demand $\boldsymbol{P}$(k) considering individual constraints and congestion prices $\boldsymbol{\Lambda}(k)$. Based on $\boldsymbol{P}$(k), the LSE establishes and optimizes the LMOS model. Then, congestion prices $\boldsymbol{\Lambda}(k)$ are obtained and broadcast to each consumer. The algorithm converges until the congestion prices change less than a given tolerance σ in consecutive iterations,

$$|\lambda^t(k+1) - \lambda^t(k)| < \sigma, \ \forall t \tag{7.44}$$

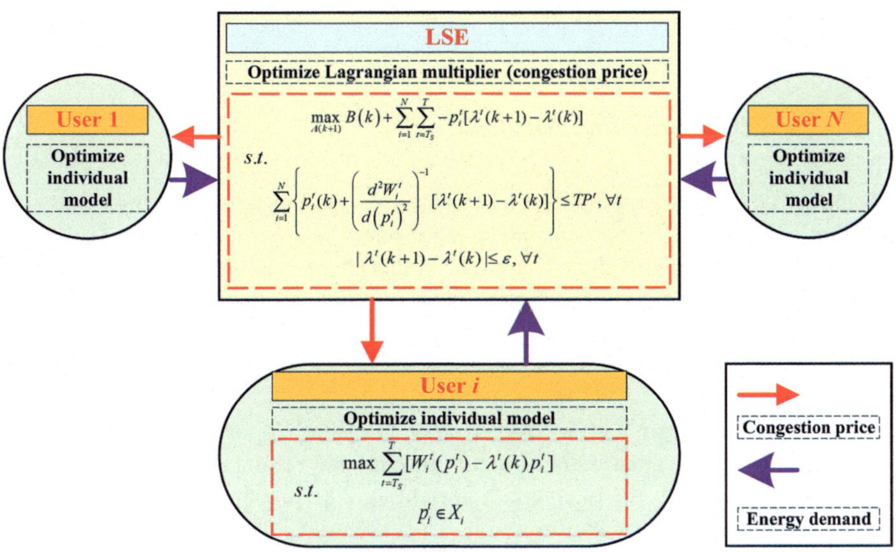

Fig. 7.3 The framework of real-time DR based on LMOS

In addition, the convergence of the LMOS approach is proved in the Appendix. Practically, the interactions between the LSE and consumers can be automatically realized by an entity or software agent such as home energy manager (HEM) at consumers' level [12, 13], which can help consumers get rid of time-consuming communication and decision-making work.

7.6 Case Studies

In the case studies, the testing environment is a Thinkpad T440p, 2.40 GHz, with 8 cores. The program is developed using MATLAB R2015a. The optimization solver is CPLEX 12.4 [32]. The RTPs in Illinois, USA on September 14, 2015 are adopted [33]. The real-time DR model is optimized by three methods: (1) M1, the distributed algorithm based on sub-gradient projection [34], (2) M2, the fast distributed algorithm in [5], and (3) the proposed distributed algorithm based on LMOS.

In M1, the dual problem is decoupled both spatially and temporally. The iterative procedure is given in eq. (7.45) and the step size $\alpha(k)$ is chosen as eq. (7.46),

$$\lambda^t(k+1) = \max\left\{\lambda^t(k) + \alpha(k) \cdot g_\lambda^t(k), 0\right\}, \ \ \forall t \tag{7.45}$$

$$\alpha(k) = \frac{1}{C_1 \cdot k + C_2} \tag{7.46}$$

where $g_\lambda^t(k)$ is a sub-gradient for the congestion price in time slot t, and C_1 and C_2 are constant parameters, the choice of which are elaborated upon in the case studies.

7.6.1 Utility Functions Analysis

The utility is a measure of preference over some goods and services, representing the satisfaction experienced by consumers. The salient feature of the utility is that the marginal utility is diminishing. Therefore, in this chapter, a linear marginal utility is adopted reflecting the diminishing satisfaction of the consumers, shown as follows:

$$MU_i = -k_i^S p_i^t + k_i^I, \ k_i^S > 0 \tag{7.47}$$

where MU_i is the marginal utility of consumer i, k_i^S is the slope, the physical meaning of which is the marginal satisfaction reduced with the increase of unit energy consumption, and k_i^I is the intercept, representing that how much marginal satisfaction is without any energy consumption. Then the utility function can be expressed as follows:

$$U_i(p_i^t) = \int_0^{p_i^t} MU_i(x)dx + k_i^C = -\frac{k_i^S}{2}(p_i^t)^2 + k_i^I p_i^t + k_i^C \tag{7.48}$$

where k_i^C is a constant, representing the satisfaction of consumer i without any energy consumption.

It is worth mentioning that there are different types of consumers with various forms of utility functions. In [35], for example, a nonlinear form of utility functions is adopted:

$$U_i^{\ln}(p_i^t) = k_i \ln(1 + p_i^t), \ k_i > 0 \tag{7.49}$$

However, this logarithm form can even be approximated as the quadratic form using Taylor expansion, i.e.,

$$\begin{aligned} U_i^{\ln}(p_i^t) &\approx U_i^{\ln}(0) + p_i^t\left[U_i^{\ln}(p_i^t)\right]'\Big|_{p_i^t=0} + (p_i^t)^2\frac{\left[U_i^{\ln}(p_i^t)\right]''\Big|_{p_i^t=0}}{2!} \\ &= -\frac{k_i}{2}(p_i^t)^2 + k_i p_i^t \end{aligned} \tag{7.50}$$

Therefore, the utility functions are set to be quadratic and concave in this chapter.

7.6.2 Smart Distribution Grid with 10 Consumers

To verify the effectiveness and efficiency of the proposed approach, a smart distribution grid with 10 consumers is studied. The time horizon is 24 and the current hour is 1.

The utility functions are quadratic and concave,

$$U_i^t(p_i^t) = \begin{cases} -a_i(p_i^t)^2 + b_i p_i^t, & 0 \le p_i^t \le \dfrac{b_i}{2a_i} \\ 0, & p_i^t > \dfrac{b_l}{2a_i} \end{cases} \tag{7.51}$$

where a_i is the opposite of the quadratic coefficient, and b_i is the monomial coefficient. The parameters of the 10 consumers are listed in Table 7.1.

Because the utility functions are only known to the consumers themselves, the LSE has to estimate the second derivatives of consumers' utility functions. The estimated values can be obtained by observing the energy demand increments in response to congestion price perturbation, shown in eq. (7.40). According to Table 7.1, the real second derivatives are $-2a_i$. The estimated and real values are

Table 7.1 Parameters of 10 consumers

Consumer	a_i ($/MW^2)	b_i ($/MW)	$\bar{p}_i^t$(MW)	$\underline{p}_i^t$(MW)	TE_i (MWh)
1	1.44	20.00	6.93	0.5	117.84
2	0.91	19.85	10.90	0.5	185.35
3	0.72	19.70	13.65	0.5	232.10
4	0.62	19.55	15.73	0.5	267.39
5	0.56	19.40	17.37	0.5	295.37
6	0.51	19.24	18.72	0.5	318.27
7	0.48	19.10	19.85	0.5	337.44
8	0.46	18.94	20.81	0.5	353.72
9	0.43	18.79	21.63	0.5	367.72
10	0.42	18.64	22.34	0.5	379.85

Table 7.2 The absolute real and estimated values of second derivatives of consumers' utility functions

Consumer	1	2	3	4	5	6	7	8	9	10
Real value	2.88	1.82	1.44	1.24	1.12	1.02	0.96	0.92	0.86	0.84
Estimated value	2.93	1.85	1.46	1.26	1.13	1.04	0.97	0.92	0.88	0.85

shown in Table 7.2. From these results, the second derivatives can be estimated accurately, which are the basis of the LMOS approach. In the following part of the case study, the LSE uses estimated second derivatives to optimize the Lagrangian multipliers while the consumers optimize their energy demands with real utility functions.

The supply capacity of the distribution grid is 125.96 MW, and the tolerance is 10^{-3} MW. The optimal congestion prices and energy demands are shown in Fig. 7.4.

From the optimal results, one can observe: (1) Energy demands are related to RTP. When RTPs are high from 12:00 to 20:00, the consumers decrease their energy demands to reduce electricity costs. (2) Congestion prices reflect the power supply shortage. From 12:00 to 20:00, the total energy demands of 10 consumers are less than the supply capacity of the distribution grid. Thus, the congestion prices during this period are zero, reflecting that the power supply is sufficient.

To test the best performance of M1, C_1 and C_2 are screened. The results of iteration times by M1 are demonstrated in Fig. 7.5.

In Fig. 7.5, the values of C_1 and C_2 greatly influence the iterations. When $C_1 = 0.20$ and $C_2 = 0.97$, M1 achieves the lowest iterations, equaling to 108.

Although the same optimum is achieved by all three methods, the fluctuations in the Lagrangian multipliers take as many as 108 and 87 iterations before stability was observed for M1 and M2. In contrast, the iteration number of M3 is only 28, which is reduced by 74.07% and 67.82% compared with those of M1 and M2, respectively.

Figure 7.6 illustrates a comparison of the congestion prices at 1:00 and the energy demand of consumer 1 at 1:00 by three methods.

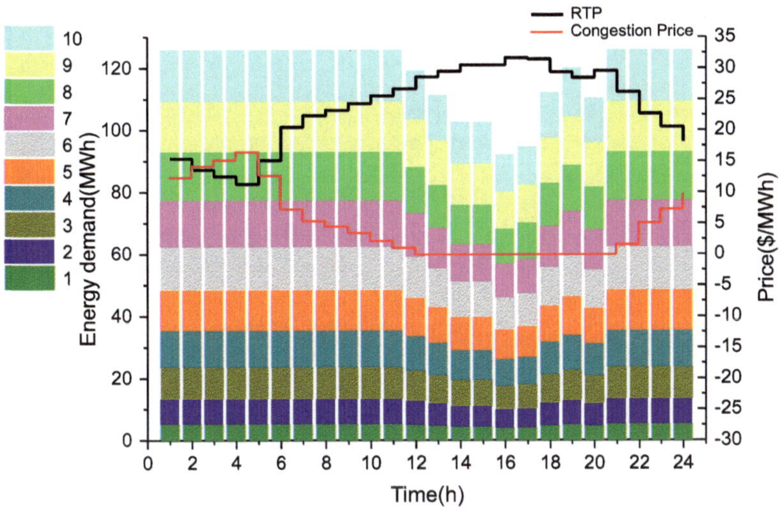

Fig. 7.4 Optimal results of distribution grid with 10 consumers

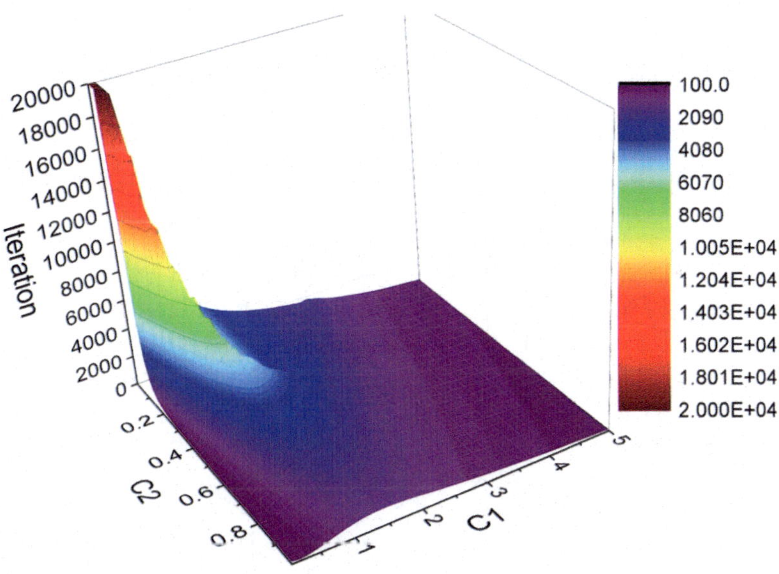

Fig. 7.5 Results of iterations by M1 screening C_1 and C_2

From the iteration results, M3 converges much faster than M1 and M2. The congestion price at 1:00 converges to 12.43 \$/MWh and the energy demand of consumer 1 at 1:00 converges to 5.19 MWh by three approaches. However, in M1, the search direction is the sub-gradient projection, and the step size is obtained in a

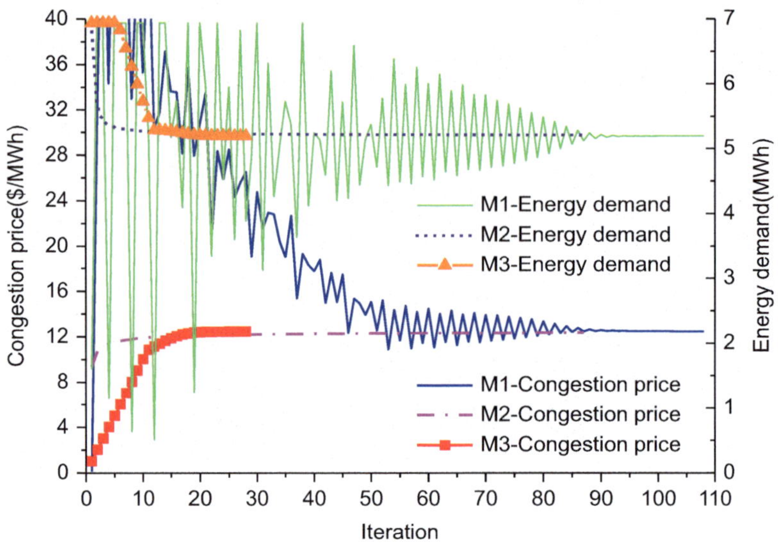

Fig. 7.6 Iteration of results by M1, M2, and M3

fixed manner, shown in eq. (7.52); in M2, although the binary search method is employed for coordination, the congestion prices still rely on sub-gradient projection, whereas in M3, the Lagrangian multipliers are optimized in a predictive way according to the iterative energy demands. Therefore, M3 converges much faster.

7.6.3 *Smart Distribution Grid with 100 Consumers*

In this case, a smart distribution grid with 100 consumers is studied. The time horizon is 24, and the current hour is 1.

The supply capacity of the distribution grid is 1705.3 MW, and the tolerance is 10^{-3} MW. The optimal congestion prices and energy demands are shown in Fig. 7.7.

From the optimal results, the energy demands of 100 consumers are less than the supply capacity from 11:00 to 21:00 due to the high RTPs. As a result, the power congestion is relieved, and the corresponding congestion prices are zero.

After screening C_1 and C_2, the minimal iterations of M1 is 416, with $C_1 = 0.30$ and $C_2 = 2.00$.

Although the same optimum is achieved by all three methods, the iterations of M3 is 34, while M1 takes 416 iterations and M2 takes 122 iterations to converge. By utilizing M3, the iterations can be reduced by 91.83% and 72.13%, compared with M1 and M2, respectively. It is demonstrated that the proposed method is effective and efficient. The convergence performance is robust and stable.

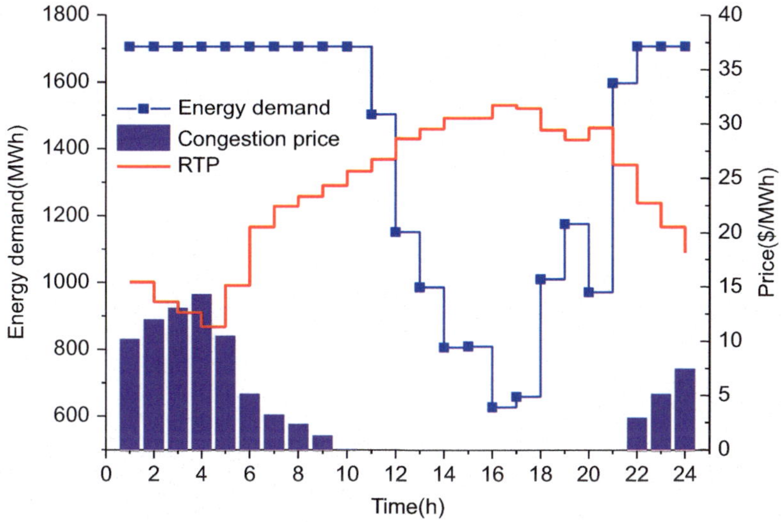

Fig. 7.7 Optimal results of distribution grid with 100 consumers

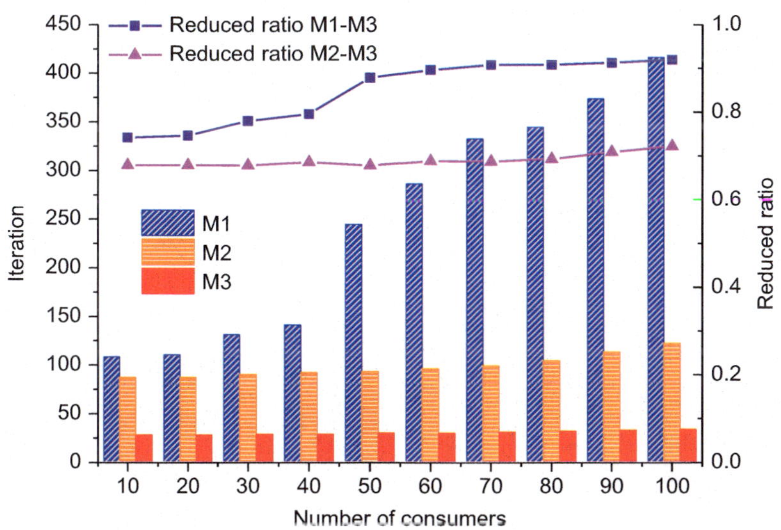

Fig. 7.8 Iterations of distribution grids with 20–100 consumers

To further verify the performance of the proposed distributed algorithm, scenarios with 20–90 consumers are simulated. In each scenario, C1 and C2 are screened to achieve the best performance of M1. The results of iterations by M1, M2, and M3 are shown in Fig. 7.8.

In Fig. 7.8, the blue bar shows the reduced ratio of iterations by M3 compared with M1. As one can observe: (1) With the increase of the consumer number, the iterations of M1 increase greatly, which reflects the critical influence of the step size and indicates the unstable convergence performance. (2) The iterations of M2 increase significantly from 87 to 122. (3) The iterations of M3 increase slightly from 28 to 34, which shows the stable convergence performance. Therefore, compared with M1 and M2, M3 achieves better efficiency and robustness, which shows the potential for real-world applications.

7.6.4 Real-World Distribution Grid with 14 Commercial Consumers

To further verify the efficiency of the proposed approach, a real-world distribution grid with 14 commercial consumers is tested. The time horizon is 24 and the current hour is 1. The energy usage, revenue, and cost data are from [36] in Illinois, USA. These commercial consumers are engaged in different business, including merchandising, manufacturing, communication, etc. The users can gain revenues by consuming electricity. Thus, the revenues are regarded as the utility of commercial consumers. The historical data and the regressed utility of consumer 1 are shown in Fig. 7.9.

Hence, the utility function of consumer 1 is,

$$U_1 = -2.863\left(p_1^t\right)^2 + 50.16 p_1^t - 14.84 \tag{7.52}$$

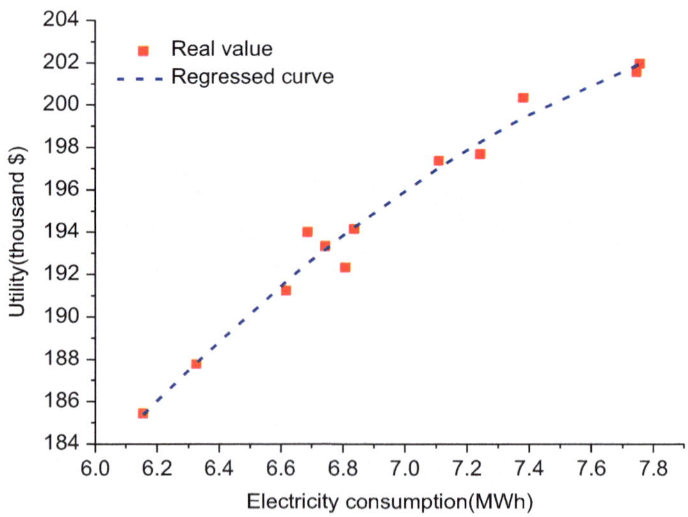

Fig. 7.9 The historical data and the regressed utility of consumer 1

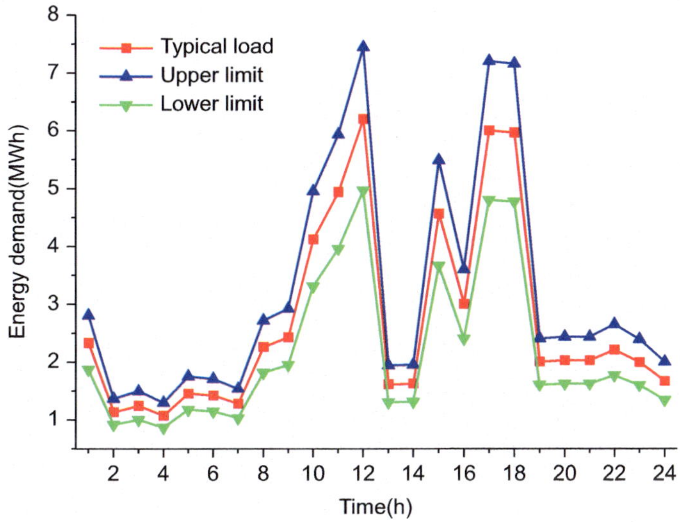

Fig. 7.10 The typical daily load demands of consumer 1

By regressing the quadratic relationship between electricity consumption and the revenues, the utility function of 14 consumers can be obtained. The typical daily load demands of consumer 1 are shown in Fig. 7.10. The upper and lower limits are set to be 1.2 and 0.8 times of the load demands.

The daily load requirements are the demands of each consumer's typical daily load. The supply capacity of the distribution grid is 60 MW. The tolerance is 10^{-3} MW. The energy demands before and after optimal scheduling are shown in Fig. 7.11a, and the optimal congestion prices are shown in Fig. 7.11b.

Restricted by the supply capacity of the distribution grid, the consumers will shift the peak loads to the valley hours to maintain the daily load requirements. When the total loads reach the supply capacity during peak hours from 9:00 to 23:00, the congestion prices are generated to leverage the balance between the energy supply and demand.

To test the best performance of M1, C_1 and C_2 are screened. When $C_1 = 0.05$ and $C_2 = 0.50$, M1 achieves the lowest iterations. Table 7.3 compares the iteration results of M1, M2, and M3. From the iteration results, M3 converges faster than M1 and M2, demonstrating the effectiveness and efficiency of the proposed method.

7.7 Conclusions

In this chapter, a real-time DR framework and model in a smart distribution grid is formulated with both spatial and temporal constraints. The objective of the model is to maximize the consumer welfare subject to the detailed constraints of each

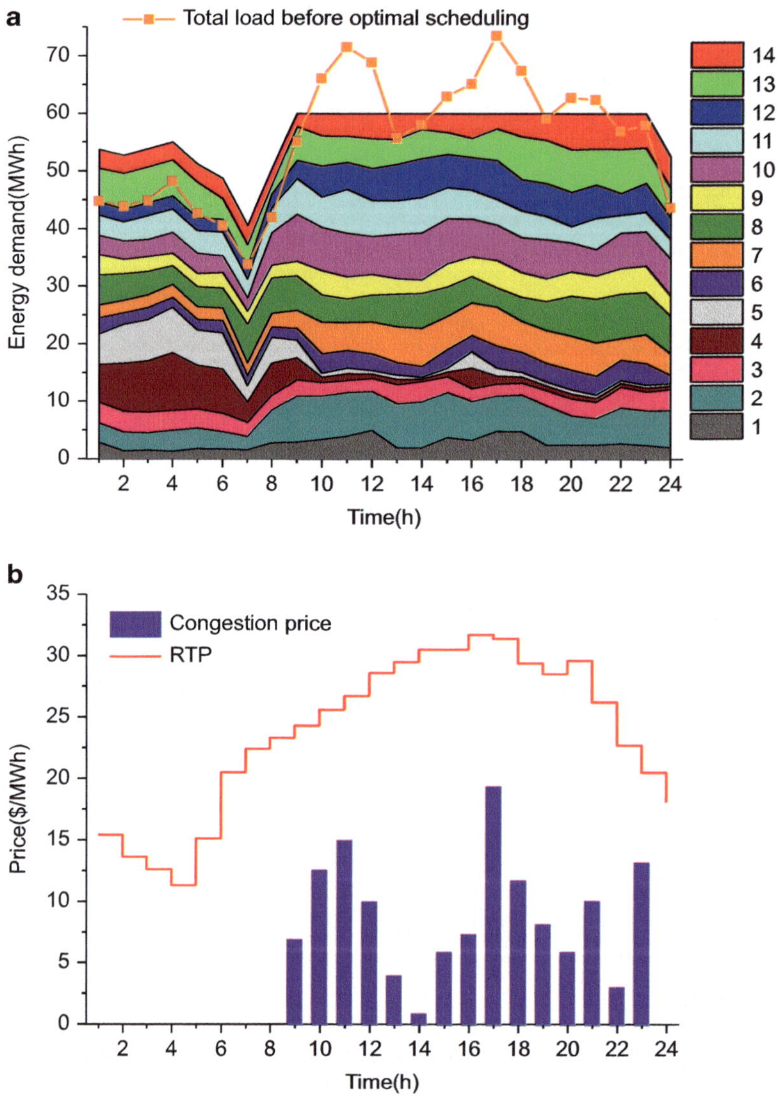

Fig. 7.11 The energy demands before and after optimal scheduling

Table 7.3 Iterations of M1, M2, and M3

Method	M1	M2	M3
Iteration	181	106	41

individual. Because the design of the iteration step size and search direction highly influences the convergence of the existing distributed algorithms, Lagrangian multiplier optimal selection is presented for the first time and applied to the real-time DR model. Using sensitivity analysis, Lagrangian multipliers are optimized within a relatively small linearized neighborhood during each iteration, greatly accelerating the convergence performance. The case studies based on a smart distribution grid with numbers of consumers ranging from 10 to 100 verify the outperformance and efficiency of the proposed algorithm. Using consumers' energy demand and demand elasticity, the proposed LMOS approach is able to provide the LSE with precise price signals to leverage the balance between the energy supply and demand. Hopefully, the proposed model and method can provide new insights for the operation of smart distribution grids with large-scale DR resources.

Appendix

Variable Sensitivity Simplification

In this subsection, the simplification of variable sensitivity is derived. According to eq. (7.38), the variable sensitivity equation is

$$d\boldsymbol{x} = \left[\boldsymbol{F}_{xx}^{-1}\boldsymbol{E}_x\left(\boldsymbol{E}_x^{\mathrm{T}}\boldsymbol{F}_{xx}^{-1}\boldsymbol{E}_x\right)^{-1}\boldsymbol{E}_x^{\mathrm{T}} - \boldsymbol{I}\right]\boldsymbol{F}_{xx}^{-1}\boldsymbol{F}_{xa}d\boldsymbol{a} \tag{7.53}$$

For the real-time DR model in this chapter, the constraints are the daily energy demands of consumer i:

$$\sum_{t=1}^{T} p_i^t \geq TE_i \tag{7.54}$$

Hence, the matrix $\boldsymbol{E}_x$ of consumer i is

$$\boldsymbol{E}_x = [1,\ 1,\ \ldots 1]^{\mathrm{T}} \in R^{(T-T_S+1)\times 1} \tag{7.55}$$

Given that

$$B_i^t\left(p_i^t, \lambda^t\right) = W_i^t\left(p_i^t\right) - \lambda^t p_i^t \tag{7.56}$$

The second derivative matrix $\boldsymbol{F}_{xx}$ of consumer i is

$$\boldsymbol{F}_{xx} = diag\left(\frac{d^2 W_i^t}{d\left(p_i^t\right)^2}\right) \tag{7.57}$$

The matrix $\boldsymbol{F}_{xa}$ of consumer i is

$$\boldsymbol{F}_{xa} = -\boldsymbol{I} \tag{7.58}$$

Therefore, the variable sensitivity equation of consumer i is

$$\begin{aligned} d\boldsymbol{p}_i &= \left[\boldsymbol{F}_{xx}^{-1}\boldsymbol{E}_x\left(\boldsymbol{E}_x^{\mathrm{T}}\boldsymbol{F}_{xx}^{-1}\boldsymbol{E}_x\right)^{-1}\boldsymbol{E}_x^{\mathrm{T}} - \boldsymbol{I}\right]\boldsymbol{F}_{xx}^{-1}\boldsymbol{F}_{xa}d\boldsymbol{\Lambda} \\ &= \left(\frac{1}{\sum_{t=T_S}^{T} Q_i^t}\begin{bmatrix} Q_i^{T_S} & \cdots & \cdots & Q_i^{T_S} \\ Q_i^{T_S+1} & \cdots & \cdots & Q_i^{T_S+1} \\ \cdots & \cdots & \cdots & \cdots \\ Q_i^T & \cdots & \cdots & Q_i^T \end{bmatrix} - \boldsymbol{I}\right)\boldsymbol{F}_{xx}^{-1}\boldsymbol{F}_{xa}d\boldsymbol{\Lambda} \\ &\approx -\boldsymbol{I}\cdot diag\left(\frac{d^2W_i^t}{d(p_i^t)^2}\right)^{-1}\cdot(-\boldsymbol{I})d\boldsymbol{\Lambda} \\ &= diag\left(1/\frac{d^2W_i^t}{d(p_i^t)^2}\right)d\boldsymbol{\Lambda} \end{aligned} \tag{7.59}$$

where Q_i^t is

$$Q_i^t = 1/\frac{d^2W_i^t}{d(p_i^t)^2} \tag{7.60}$$

Convergence Analysis

In this subsection, the convergence of the proposed LMOS approach is proved.

Monotone Convergence Theorem [37]:

If $\{a_n\}$ is a monotone sequence of real numbers, i.e.,

1. if $a_n \le a_{n+1}$ for every $n \ge 1$; or,
2. if $a_n \ge a_{n+1}$ for every $n \ge 1$,

then this sequence has a finite limit if and only if the sequence is bounded.

The model of LSE for optimally determining the congestion prices in next iteration is listed as follows:

$$
\begin{aligned}
& B^*(k+1) = \max_{\Lambda(k+1)} \sum_{i=1}^{N} \sum_{t=T_S}^{T} \left(B_i^t(k) + dB_i^t\right) \\
& \text{s.t.} \\
& \sum_{i=1}^{N} \left\{ p_i^t(k) + \left[\frac{d^2 W_i^t}{d(p_i^t)^2}\right]^{-1} [\lambda^t(k+1) - \lambda^t(k)] \right\} \le TP^t, \ \forall t \\
& |\lambda^t(k+1) - \lambda^t(k)| \le \varepsilon, \ \forall t
\end{aligned}
\tag{7.61}
$$

where $B^*(k + 1)$ represents the objective of the model in the $(k + 1)^{\text{st}}$ iteration, and dB_i^t is the function of $\Lambda(k + 1)$, i.e.,

$$
dB_i^t = -p_i^t(k) \cdot [\lambda^t(k+1) - \lambda^t(k)], \ \forall i, t \tag{7.62}
$$

Then the following expression holds:

$$
\begin{aligned}
B^*(k+1) &= \max_{\Lambda(k+1)} \sum_{i=1}^{N} \sum_{t=T_S}^{T} \left(B_i^t(k) + dB_i^t\right) \\
&= B^*(k) + \max_{\Lambda(k+1)} \sum_{i=1}^{N} \sum_{t=T_S}^{T} dB_i^t \ge B^*(k)
\end{aligned}
\tag{7.63}
$$

Then, the sequence $\{B^*(k)\}$ is monotonically increasing, while it is obvious that the sequence is upper bounded because of consumers' finite utility. Therefore, according to *Monotone Convergence Theorem*, the sequence $\{B^*(k)\}$ must converge with the increase of the index k.

To further verify the optimality of the converged solutions, the property of the sub-gradient projection method should be employed: If there exists a valid Lipschitz constant for the dual function, the distributed algorithm can converge to the optimum for a sufficiently small step size. The details can be widely found in the existing literature [5, 38].

References

1. Siano, P., & Sarno, D. (2016). Assessing the benefits of residential demand response in a real time distribution energy market. *Applied Energy, 161*, 533–551.
2. Yu, Z., Chen, S., & Tong, L. (2016). An intelligent energy management system for large-scale charging of electric vehicles. *CSEE Journal of Power and Energy Systems, 2*(1), 47–53.
3. Li, Z., Wu, W., Zhang, B., et al. (2013). Dynamic economic dispatch using Lagrangian relaxation with multiplier updates based on a quasi-Newton method. *IEEE Transactions on Power Apparatus and Systems, 28*(4), 4516–4527.
4. Lai, X., Xie, L., Xia, Q., et al. (2014). Decentralized multi-area economic dispatch via dynamic multiplier-based Lagrangian relaxation. *IEEE Transactions on Power Apparatus and Systems, 30*(6), 3225–3233.

5. Deng, R., Xiao, G., Lu, R., et al. (2015). Fast distributed demand response with spatially- and temporally- coupled constraints in smart grid. *IEEE Transactions on Industrial Informatics, 11* (6), 1597–1606.
6. Xing, H., Cheng, H., & Zhang, L. (2015). Demand response based and wind farm integrated economic dispatch. *CSEE Journal of Power and Energy Systems, 1*(4), 47–51.
7. Arteconi, A., Patteeuw, D., Bruninx, K., et al. (2016). Active demand response with electric heating systems: Impact of market penetration. *Applied Energy, 177*, 636–648.
8. Nolan, S., & O'Malley, M. (2015). Challenges and barriers to demand response deployment and evaluation. *Applied Energy, 152*, 1–10.
9. Sheikhi, A., Rayati, M., Bahrami, S., et al. (2015). Integrated demand side management game in smart energy hubs. *IEEE Transactions on Smart Grid, 6*(2), 675–683.
10. Wang, Q., Zhang, C., Ding, Y., et al. (2015). Review of real-time electricity markets for integrating distributed energy resources and demand response. *Applied Energy, 138*, 695–706.
11. Chen, S., & Liu, C. (2017). From demand response to transactive energy: State-of-the-art. *Journal of Modern Power Systems and Clean Energy, 5*(1), 10–19.
12. Papadaskalopoulos, D., & Strbac, G. (2013). Decentralized participation of flexible demand in electricity markets-part I: Market mechanism. *IEEE Transactions on Power Apparatus and Systems, 28*(4), 3658–3666.
13. Fan, Z. (2012). A distributed demand response algorithm and its application to PHEV charging in smart grids. *IEEE Transactions on Smart Grid, 3*(3), 1280–1290.
14. Zhong, H., Xie, L., Xia, Q., et al. (2013). Coupon incentive-based demand response: Theory and case study. *IEEE Transactions on Power Apparatus and Systems, 28*(2), 1266–1276.
15. Finn, P., & Fitzpatrick, C. (2014). Demand side management of industrial electricity consumption: Promoting the use of renewable energy through real-time pricing. *Applied Energy, 113*, 11–21.
16. Behboodi, S., Chassin, D., Crawford, C., et al. (2016). Renewable resources portfolio optimization in the presence of demand response. *Applied Energy, 162*, 139–148.
17. Deng, R., Yang, Z., Hou, F., et al. (2015). Distributed real-time demand response in multiseller-multibuyer smart distribution grid. *IEEE Transactions on Power Apparatus and Systems, 30*(5), 2364–2374.
18. Chen, C., Wang, J., & Kishore, S. (2014). A distributed direct load control approach for large-scale residential demand response. *IEEE Transactions on Power Apparatus and Systems, 29*(5), 2219–2228.
19. Safdarian, A., Firruzabad, M., & Lehtonen, M. (2014). A distributed algorithm for managing residential demand response in smart grids. *IEEE Transactions on Industrial Informatics, 10*(4), 2385–2393.
20. Mazidi, M., Monsef, H., & Siano, P. (2016). Robust day-ahead scheduling of smart distribution networks considering demand response programs. *Applied Energy, 178*, 929–942.
21. Ongsakul, W., & Petcharaks, N. (2004). Unit commitment by enhanced adaptive Lagrangian relaxation. *IEEE Transactions on Power Apparatus and Systems, 19*(1), 620–628.
22. Conejo, A., Morales, J., & Baringo, L. (2010). Real-time demand response model. *IEEE Transactions on Smart Grid, 1*(3), 236–242.
23. Forushani, E., Moghaddam, M., Sheikh-El-Eslami, M., et al. (2014). Risk- constrained offering strategy of wind power producers considering intraday demand response exchange. *IEEE Transactions on Sustainable Energy, 5*(4), 1036–1047.
24. Kazemi, M., Mohammadi-Ivatloo, B., & Ehsan, M. (2015). Risk-constrained strategic bidding of gencos considering demand response. *IEEE Transactions on Power Apparatus and Systems, 30*(1), 376–384.
25. Wei, W., Liu, F., & Mei, S. (2015). Energy pricing and dispatch for smart grid retailers under demand response and market price uncertainty. *IEEE Transactions on Smart Grid, 6*(3), 1364–1374.

26. Epstein, L. G., & Zin, S. E. (1991). Substitution, risk aversion, and the temporal behavior of consumption and asset returns: An empirical analysis. *Journal of Political Economy, 99*, 263–286.
27. Gountis, P. V., & Bakirtzis, G. A. (2004). Bidding strategies for electricity producers in a competitive electricity marketplace. *IEEE Transactions on Power Apparatus and Systems, 19* (1), 356–365.
28. Fahrioglu, M., & Alvarado, L. F. (2001). Using utility information to calibrate customer demand management behavior models. *IEEE Transactions on Power Apparatus and Systems, 16*(2), 317–322.
29. Harrison, G. W. (2008). *Maximum likelihood estimation of utility functions using Stata*. University of Central Florida, Working Chapter. 6–12.
30. Ratliff, L. J., Dong, R., Ohlsson, H., & Sastry, S. S. (2014). Incentive design and utility learning via energy disaggregation. *IFAC Proceedings Volumes, 47*(3), 3158–3163.
31. Li, F., & Bo, R. (2007). DCOPF-based LMP simulation: Algorithm, comparison with ACOPF, and sensitivity. *IEEE Transactions on Power Apparatus and Systems, 22*(4), 1475–1485.
32. The IBM ILOG CPLEX website. 2015. Retrieved from http://www-01.ibm.com/software/websphere/products/optimization/academic-initiative/index.html/.
33. Ameren Illinois Power Corporation, Real time prices for residential customers, 2015. Retrieved from https://www2.ameren.com/RetailEnergy/RealTimePrices
34. Zhu, M., & Martinez, S. (2012). On distributed convex optimization under inequality and equality constraints. *IEEE Transactions on Automatic Control, 57*(1), 151–164.
35. Roh, H., & Lee, J. (2016). Residential demand response scheduling with multiclass appliances in the smart grid. *IEEE Transactions on Smart Grid, 7*(1), 94–104.
36. U.E.I. Administration 2015. Retrieved from https://www.eia.gov/electricity/data.cfm#sales
37. Bibby, J. (1974). Axiomatisations of the average and a further generalisation of monotonic sequences. *Glasgow Mathematical Journal, 15*(1), 63–65.
38. Bertsekas D. (1999). *Nonlinear programming*. Athena Scientific.

Chapter 8
Load Resources to Provide Primary Frequency Reserve Service

8.1 Introduction of Frequency Control

Maintaining system frequency at its target value is critical to power grid operation [1–4]. The overall task of controlling frequency is organized in three levels in North America, namely, the primary, secondary, and tertiary frequency control.

The primary frequency control is the actions provided by the interconnection to arrest and stabilize frequency in response to frequency deviations. Primary control comes from automatic generator governor response, load response (typically from motors), and other devices that provide an immediate response based on local (device-level) control systems. Primary frequency response is designed for severe contingencies such as large generators tripping off-line. It involves an automatic and quick action of a generator to change its output in response to a large frequency deviation (within seconds).

Secondary frequency control is the actions provided by an individual Balancing Authority or its Reserve Sharing Group to correct the resource-load unbalance that created the original frequency deviation, which will restore both scheduled frequency and primary frequency response. Secondary control comes from either manual or automated dispatch from a centralized control system. Secondary frequency control is designed for small frequency deviations and delivers a slow response. It is activated by Automatic Generation Control (AGC) to provide an automatic response. It is able to maintain system frequency within the boundary when there is a small demand change.

Tertiary frequency control is the actions provided by Balancing Authorities on a balanced basis that are coordinated, so there is a net-zero effect on area control error (ACE). Examples of tertiary control include dispatching generation to serve native load, economic dispatch, dispatching generation to affect interchange, and re-dispatching generation. Tertiary control actions are intended to replace secondary control response by reconfiguring reserves. Tertiary control operates on a longer time scale (minutes) than primary and secondary frequency control. Its main purpose

P. Du et al., *Demand Response in Smart Grids*,
https://doi.org/10.1007/978-3-030-19769-8_8

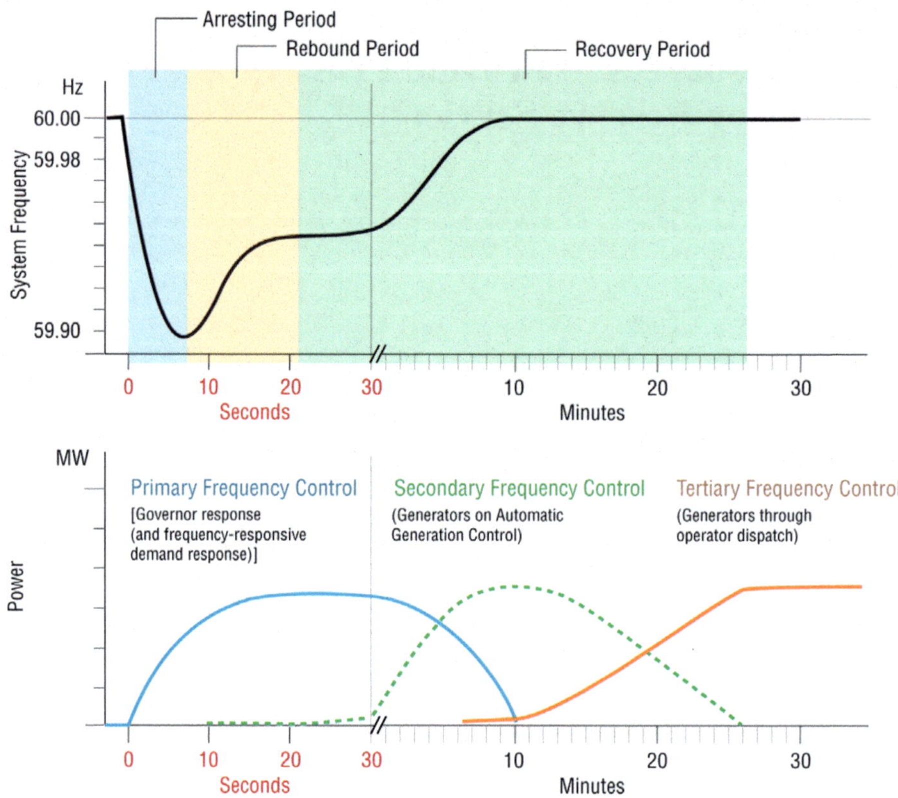

Fig. 8.1 Primary/secondary/tertiary frequency control characters and time line (source: [4])

is to replenish reserves that have been used to provide primary and secondary frequency control so that the system can be prepared for the next event.

These three stages of control work in a complementary fashion to arrest, stabilize, and restore system frequency in response to energy and load imbalance. Immediately following a loss of generation event, all synchronous rotating machines in the system provide inertial responses. During this stage, all synchronous generators supply extra active power in proportion to their relative size. Primary frequency response then kicks in and seeks to arrest and stabilize frequency before shedding any portion of firm load. This stage is known as the "arresting period" in Fig. 8.1. Once system frequency stops declining, primary frequency control will stabilize it by continuing delivering response. After frequency being stabilized, the secondary frequency control takes action to restore system frequency back to normal operational range. Eventually, tertiary control will restore the reserved capacity that have provided primary frequency control by dispatching other generators and free up the reserve.

8.2 Impact of Renewable Resource over Inertia and Primary Frequency Control

Future power systems will see significant growth of renewable generation resources (RGSs). In regions such as Ireland and Texas, power generated by RGSs has exceeded 40% of the system load. In many regional grids, fossil fuel units (e.g., coal power plants) are quickly being displaced by inertial-less RGSs (primarily wind and solar), which makes the frequency control a demanding work. In [3–6], the authors demonstrate that the large-scale integration of RGSs leads to decline in system inertia, causing a significant reduction of the primary frequency control (PFC) capability. Electric grids in North America have witnessed this deterioration in the PFC capability over the past decades. Recently, the deterioration is accelerating because the large-scale adoption of wind and solar energies is becoming a reality. Similar studies conducted for the Ireland grid also show that if no grid enhancement is implemented soon, the risk of constraining or curtailing significant amount of wind generation would be very high because of the lack of adequate PFC [7, 8]. Therefore, whether or not those low-inertia interconnected power systems can procure and maintain adequate frequency response reserve (FRR) for providing PFC to respond to credible contingencies is becoming a serious concern [9, 27–31].

Alternative resources for providing FRR include synthetic inertia from wind turbines, energy storage, and demand response. Synthetic inertia from wind turbines cannot sustain the response unless a portion of wind power is spilled [10, 11]. Fast-acting energy storage systems such as batteries or flywheels can quickly stabilize the grid frequency after a disturbance is detected but their costs are still prohibitively high [12]. Traditionally, demand response is used mainly for peak shaving and load shifting. Recent advancements in communication, sensor, and automation technologies allow fast responsive load to provide fast, high-quality grid services such as load following and regulation [13]. Simulation results and pilot projects have shown that commercial building and residential home loads can effectively respond to under-frequency events such as the loss of large generators [14, 15]. Another advantage for demand response to be frequency responsive is its low cost in the provision of service, which only requires a local device installed to detect frequency deviations and to interrupt the load when needed. Thus, introducing fast-acting load-side resources for FRR is both technically possible and economically efficient.

However, existing studies [16–23] on primary frequency market design, which have built a framework for hosting services provided only by generation resources, did not consider the role of demand response in such a market. The basic principles of co-optimizing energy and primary, secondary, and tertiary reserves are discussed in [21]. In [16, 17], the authors propose a market design mechanism for incentivizing synchronous generators to provide primary frequency response. A simplified dynamic model is introduced to determine the minimum spinning reserve requirement that is used as part of the constraints in economic dispatch for a pool-based power market [18–20, 22]. The problem formulation of accounting for primary service constraints in unit commitment is described in [23]. Those studies focused primarily on the provision of primary frequency response from synchronous generators without considering the participation of other viable resources.

The participation of load resources in the ancillary service market is promising as it can improve both reliability and efficiency. However, there are several challenges associated with the integration of load resources into the primary frequency reserve market. The main difficulty lies in the different performance among a variety of resources in the provision of the primary frequency response. The governor responses from the synchronous machines are continuous and slow, while the actions from load resources in the response to large frequency deviations are discrete and fast. In addition, when introducing the load-provided FRR to the PFC market, it is desirable to know the optimal redistribution of the FRR requirement between loads and generators and to understand how the energy and ancillary service market prices will be affected once loads can bid into both markets. To this end, dramatic changes to existing market mechanism are needed to ensure that a primary frequency reserve market is open to competition, efficient and nondiscriminatory.

Without the changes required, the existing market design can be a barrier to accommodating load-side resources for providing FRR. First, in some regions, e.g., in Western Electricity Coordinating Council, the provision of FRR is still mandatory, so there is little monetary incentive for other resources to provide such services [3]. Second, different resources have different response speeds and the sustainability in maintaining the responses are also different [14, 15]. Thus, the contribution from each resource for stabilizing the frequency excursions needs to be quantified so that different resources with different performance can be harmonized. Third, the most efficient resources should be awarded most so that the market can send an incentive price signal to eventually improve the market efficiency. This is essential when designing a market mechanism for the provision of FRR such that the load and synchronous machines can compete fairly in a single market and be compensated based on their performance.

Therefore, the new market design should focus on addressing two key issues for accommodating FRR provided by multiple resources at the PFC market: (1) to quantify the effectiveness of fast-response load resources in comparison to generation resources and (2) to design a new PFC market mechanism to reward diversified resources based on their performance in the PFC provision. The rest of this chapter will first introduce the mathematical formulation of a novel primary frequency reserve market where the load resources can participate, and then demonstrate the merits and implications of this market through a case study.

8.3 Frequency Response Reserves at ERCOT

The Electric Reliability Council of Texas (ERCOT) is an Independent System Operator (ISO) serving over 23 million customers in Texas. As a single Balancing Authority (BA) without synchronous connections to its neighboring systems, ERCOT relies purely on its internal resources to balance power shortages and variations.

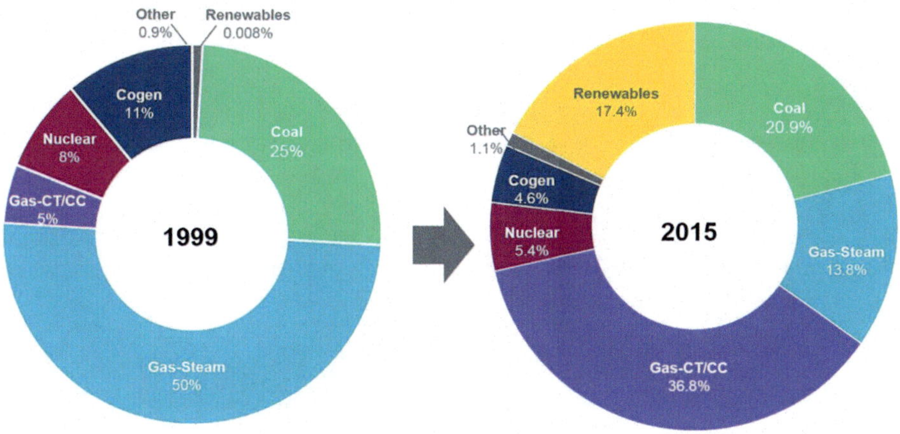

Fig. 8.2 ERCOT generation mix change

In Texas, the leading state in wind installed capacity, a total of over 20 GW wind generation is currently being operated by the state's grid operator ERCOT. The all-time highest instantaneous wind penetration record is roughly 50%. The increase in wind installed capacity at ERCOT is also accompanied by a significant change in generation mix over the past decade (Fig. 8.2). Conventional gas-steam generators are replaced with the cleaner renewables and more flexible combine cycles and combustion turbines.

Inertial energy is stored in the rotating elements of the power system. The inertia of a generator refers to the generator's resistance to the change of its rotation speed. Conventional generators with rotating turbines that are synchronously connected to the power grid store kinetic energy when they are spinning online. Immediately after a sudden loss of generation or increase in load, the stored kinetic energy will be automatically released from the synchronized machines to make up for the energy deficiency. During the process of the energy being released, the synchronized machines have to slow down their rotating turbines, so that, as a consequence, system frequency will start falling. The frequency decline of a system with high inertia will be slower than that of a lower inertia system. Therefore, inertia provides crucial values to system reliability in arresting the frequency changes.

One of the main impacts over the grid operation of renewables when replacing conventional synchronous generators is the reduced inertia [28–31]. Due to the nature of how renewable energy resources such as wind and solar are integrated into the grid, through power electronics devices, they do not contribute to overall system inertia. As more renewables are being integrated, the future power systems are expected to be operated in lower inertia conditions than ever before. Figure 8.3 is how operation inertia was correlated with respect to hourly wind penetration at ERCOT. At a high penetration of wind generation, system inertia was significantly low. A low inertia system is more sensitive to the disturbance and often is accompanied with large frequency excursions even other operation conditions (reserve) are identical or similar.

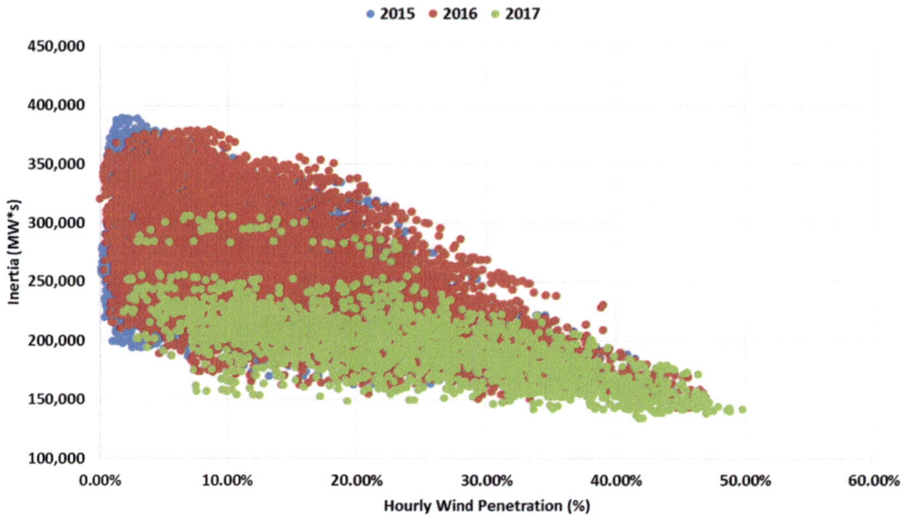

Fig. 8.3 Wind penetration vs. Inertia in ERCOT

Table 8.1 ERCOT ancillary service products

ERCOT AS products		Description
Regulation up/down		To respond within 5 s to meet real-time load
Frequency response reserve (FRR)	PFR	To provide primary frequency control (PFC)
	FFR	
Non-spinning		To cover for uncertainties in load and wind forecast and to restore reserves

To provide frequency control, reserve need to be allocated, which is simply headroom or unloaded capacity. ERCOT offers three ancillary service (AS) products: Regulation up/down, FRR, and Non-Spinning, as shown in Table 8.1.

Regulation reserve mostly provides secondary frequency control [27, 29]. It is used to balance small imbalance between generation and load in real-time operation. It is provided by resources that are able to respond to AGC signals. In most of the independent system operators (ISOs) in the USA, grid operators send out base point dispatch signal to resources every 5 min to balance the system with the most updated knowledge of where load is heading. Small variations of actual demand to the base point are automatically covered by regulation reserves. Regulation reserve is bidirectional. When demand exceeds base point, regulation up is deployed and when demand is lower than base point, regulation down takes control.

In comparison, frequency response reserve (FRR) can be quickly deployed within seconds to protect the grid when the frequency deviations are large [28, 30, 31]. FRR (or spinning reserve) is intended to quickly arrest frequency decline due to forced outages or other contingencies where generation and load are significantly imbalanced all of sudden. Establishing a PFC market, which also incentivizes load-side participation of the FRR service, will make the AS market more competitive and efficient [26]. Therefore, the proposed FRR service is extended to two categories: primary frequency response (PFR) provided by synchronous generators and fast frequency response (FFR) provided by other resources (e.g., fast response load and energy storage systems).

PFR is provided by online synchronous generators through governor response or governor-like actions to arrest and/or counter-respond to frequency deviations. As a single BA, ERCOT must comply with the BAL-003 standard [24]. The frequency response obligation (FRO) for ERCOT is 413 MW/0.1 Hz. To meet this requirement, ERCOT requires every resource with a speed governor to put the governor in service whenever the resource is online. In addition, the droop setting should not exceed 5% and the frequency response dead band should be no more than ±0.018 Hz.

FFR is a response from a resource that is self-deployed to provide a full response within 30 cycles after frequency meets or drops below certain threshold [25]. To provide FFR, a load resource will be equipped with an under-frequency relay (e.g., triggered if the frequency is dropped under 59.7 Hz). As required by ERCOT, the response time of FFR should be less than 500 ms (including the frequency relay pickup delay and the breaker action time). This makes FFR more effective, when inertia is low and rate of change of frequency is high, to mitigate the decline of frequency compared to PFR because a generator needs a few seconds to fully deliver its reserved capacity for frequency response. It is also able to improve the frequency nadir and be instrumental in preventing frequency from dropping below the involuntary under-frequency load shedding (UFLS) threshold when losing large generation units.

In the past, the minimum amount of FRR needed for PFC was determined by studying the worst-case contingency scenario. For example, a minimum of 2800 MW FRR is required for every hour in a year [9, 25] at ERCOT. This amount is calculated by the worst-case scenario corresponding to the simultaneous loss of two largest generation units. However, integrating large amounts of nonsynchronous generation resources has caused a noticeable decline in system inertia. As a result, in a low-inertia hour, faster frequency decay (i.e., higher frequency change rate) and lower frequency nadir can occur when certain generators have forced-outage events. Therefore, using the worst-cast contingency scenario (i.e., loss of two largest generation units) to determine the FRR needs becomes either inefficient (e.g., over-procure FRR at high-inertia hours) or inadequate (e.g., under-procure FRR at low-inertia hours). Thus, a methodology has to be developed to quantify the minimum FRR requirement on an hourly basis for meeting reliability requirements.

8.4 Quantification of FRR Requirement

To decide precisely how much reserve is needed is the prerequisite to the successful market operation where the reliability and economics are balanced. When the FRR requirement is concerned, the key influencing factor is system inertia, which also varies with respect to system loading conditions. In contrast to current practice where the size of reserve is calculated from a worst scenario, it is desirable to link the FRR requirement with the system inertia. In this way, the impact of anticipated operational conditions (inertia) can be reflected in the FRR quantities.

From the security perspective, the adequate amount of FRR should be allocated to ensure that, for contingencies such as the loss of two largest units, the system frequency should be arrested before triggering UFLS and the frequency nadir should be maintained above 59.3 Hz. In other words, the impact of the loss of two largest generation units can be mitigated efficiently by procuring different combinations of FFR and PFR. The following procedures are used to determine the minimum PFR and FFR requirement for different system inertias.

8.4.1 Step 1: Selection of Representative Operation Conditions

An analysis of annual operation data is conducted to cluster system inertia for ten representative net-load levels ranging from 15 to 65 GW, as shown in Fig. 8.4. At each net-load level, the medium inertia condition is selected as one of the base cases for conducting the dynamic simulation to determine the FRR requirement. Apart from the ten cases, two extreme cases, the maximum and the minimum inertia conditions, are also chosen. Collectively, twelve cases represent the full range of system inertia conditions for determining FRR need.

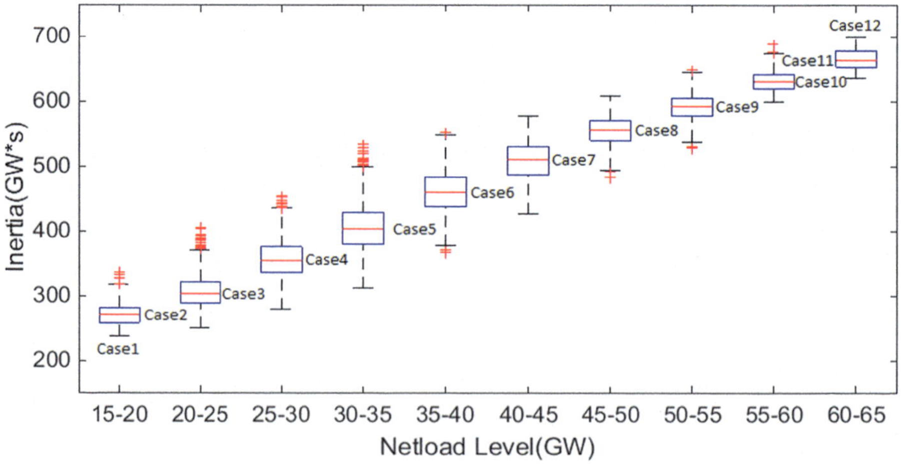

Fig. 8.4 Representative cases for system inertia selection

8.4.2 Step 2: Setup of Dynamic Models

To accurately capture the frequency response of ERCOT system, dynamic models were used with power flow data retrieved from the ERCOT dynamic security assessment databases. These are the same models which are normally used by grid operators to conduct online transient stability analysis for determining the transfer limits across the major transmission lines. The databases include detailed dynamic models of generation units and interruptible load resources. All selected governors are assumed to be operated with 5% droop setting. The triggering frequency of the FFR is set at 59.7 Hz. The pickup delay of the relay is 0.33 s and the breaker action time is 0.083 s. Standard ZIP models (constant impedance, constant current, and constant power) are used. The load damping factor is 2%/Hz. Load frequency model is used to represent the sensitivity of constant power and constant current load components to bus frequency. The first stage of the firm load UFLS program is 59.3 Hz, so the under-frequency limit for this study is set at 59.4 Hz to allow a margin of 0.1 Hz. The loss of two largest generation units (2750 MW) is simulated to assess the system frequency response. The frequency response criterion is in compliance with NERC reliability standard, which is given as "The system frequency shall be arrested between 59.4 and 60.4 Hz so that other generators will not cascade into out-of-step conditions and the firm load UFLS will not be activated".

8.4.3 Step 3: Quantification of Minimum FRR Requirement

In the simulation, the minimum FRR requirement is quantified for a given system inertia condition, and then the same procedure is repeated for different levels of the system inertia. For one of twelve cases, the system inertia is maintained the same by not changing the commitment status for the generation units. Then, the operation condition is modified by including or removing the governor control or connecting or disconnecting the dynamic models of the interruptible loads. The minimum requirement of FRR for this inertia condition can be identified when the lowest frequency dip equals to 59.4 Hz after two largest units (2750 MW) are taken offline. One example of the frequency response obtained from dynamic simulations is shown in Fig. 8.5. In this case with a system inertia at 354 GW·s, the system frequency is arrested at exactly 59.4 Hz after two largest generation units are tripped. Thus, the minimum FRR requirement is 1240 MW of PFR in addition to 1450 MW of FFR. Through a large number of simulations, a family of the curves can be derived to reveal how the minimum FRR requirements change when the combination of PFR/FRR differs. Figure 8.6 depicts three curves representing the low, medium, and high inertia conditions.

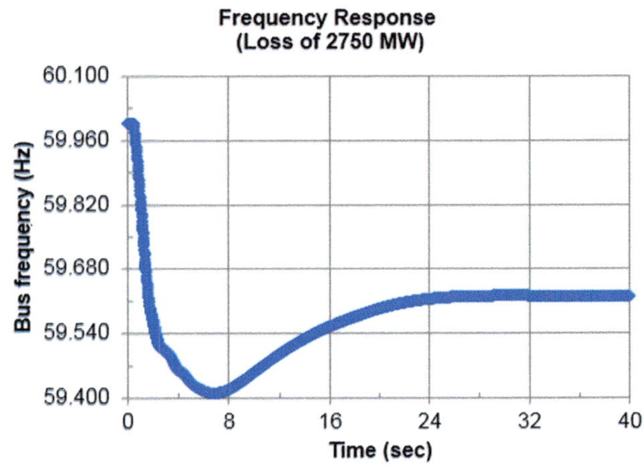

Fig. 8.5 Simulated frequency response when the system inertia is 354 GW·s

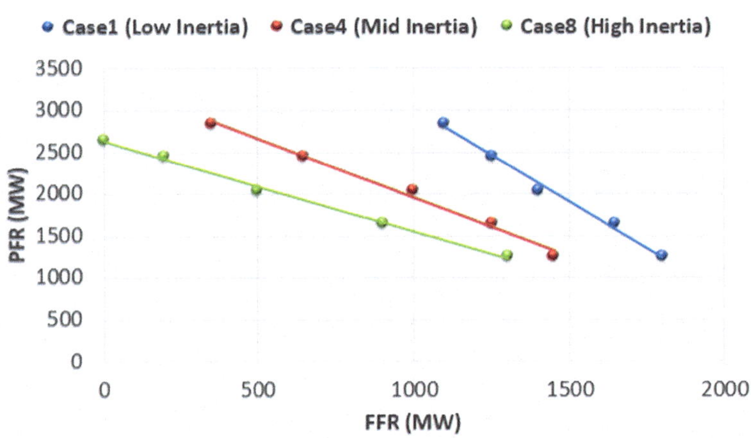

Fig. 8.6 Combinations of PFR and FFR for meeting the minimum FRR requirements at low, medium, and high system inertia conditions

8.4.4 Step 4: Derivation of Equivalency Ratio for All Cases

Through the experiments, it is observed that when the amount of FFR increases, the required amount of PFR will be reduced to protect the system integrity. As the relationship between PFR and FFR requirements is approximately linear, the slope of the lines can be viewed as an equivalency ratio, *m*, between PFR and FFR requirements. In this particular example, *m* is 1.4 and the minimum total FRR requirement is 3370 MW, so that the constraint can be described by

$$1.4 \cdot P_{FFR} + P_{PFR} \geq 3370 \tag{8.1}$$

It means that 1 MW of FFR is 1.4 times as effective as 1 MW of PFR in arresting the frequency decline. Alternatively, each 1 MW of FFR can be replaced by 1.4 MW of PFR when the system inertia condition is at 354 GW·s.

Table 8.2 summarizes the results of all 12 cases for the ERCOT grid. In general, FFR is more effective than PFR when arresting the frequency decline at low inertia conditions (e.g., Cases 1–6). However, the effectiveness decreases when the system inertia increases. When inertia is higher than 593 GW·s, FFR and PFR become equally effective. This is because there are more generators online under heavy loading conditions so that the overall system inertia is higher, making the rate of change of frequency following the disturbance much less than that in a low-inertia condition. As a result, aggregated slow-acting governor-like response has enough time to react to the loss of generation. The results also demonstrate that fast load response is more valuable in arresting the frequency drops when response speed is more critical, which is the case in low-inertia systems.

The historical equivalency ratios for the ERCOT grid in 1 year are then calculated. As shown in Fig. 8.7, in hot summer peak load hours (noon to evening), m equals to 1.0. For early morning hours in shoulder and winter months, m can be as high as 2.2.

Once the FRR requirement is determined as shown in Table 8.2, it can be included into the market clearing process. Essentially, the requirement of FFR can be mathematically described as a function of system inertia, the equivalency ratio, m, and the amount of PFR available. The equivalency ratio, m, is the performance metric for FFR and allows PFR to be substituted by FFR without compromising the system primary frequency control capability. Thus, the minimum amount of the FRR requirement becomes

$$\begin{aligned} & m \cdot P_{FFR} + P_{PFR} \geq FRR_{\min} \\ & P_{PFR} \geq PFR_{\min} \end{aligned} \tag{8.2}$$

where

Table 8.2 Minimum FFR requirement and equivalency ratios

Case No.	Inertia (GW·s)	$FRR_{\min}$(MW)	Equivalency ratio (m)
1	239	5200	2.2
2	271	4700	2.0
3	304	3750	1.5
4	354	3370	1.4
5	403	3100	1.3
6	459	3040	1.25
7	511	2640	1.13
8	556	2640	1.08
9	593	2240	1
10	631	2280	1
11	664	2140	1
12	700	2140	1

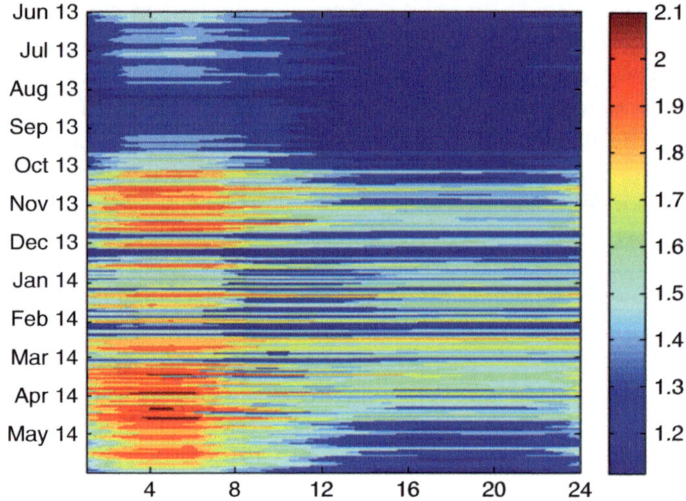

Fig. 8.7 Heat map of historical equivalency ratios between PFR and FFR

$$m = \frac{P_{PFR}}{P_{FFR}}$$

As indicated above, the equivalency ratio, *m*, explicitly accounts for the different performance between FFR and PFR, and the formulation modeled does not limit the provision of FRR service to PFR only.

8.5 Co-optimization of Energy and FRR in Day-ahead Market

8.5.1 Day-ahead market co-optimization model

A new day-ahead energy, inertia, and reserve co-optimization formulation are proposed in which the FRR requirement can be met by both primary frequency reserve (PFR) from synchronous generators and fast response reserve (FFR) from load. The proposed work reported is an extension of the previous work which studied the day-ahead market co-optimizing energy and reserves [32–38]. Compared to [31], this formulation can explicitly consider the interdependency between the commitment decision, inertia, and PFR/FFR. Unlike the study in [39], the work conducted here determines the award of FFR in the framework of the unit commitment while it does not require the dynamic simulation as part of the optimization problem, which makes the problem more tractable.

The day-ahead market energy and reserve co-optimization problem is a security-constrained unit commitment problem. The objective function is to maximize the social welfare which is the sum of demand benefits based on demand bids minus energy costs based on generators' three part offers, reserve costs based on reserve offers from both generating units and demands, and unserved reserve cost based on penalty prices as shown in (8.3). The start-up cost is a function of the number of hours the generator has been turned off. The constraint (8.4) implicitly determines the start-up costs of generating units in each hour during the optimization process. q represents the index of time segments in stepwise start-up curves. STC_{it} will be zero resulted from the optimization process if the unit does not change its status from off to on at hour t.

Formulations (8.5)–(8.8) represent unit commitment status coupling constraints. The start-up and shutdown indicators are relaxed in this study to be continuous variables in (8.8) in order to accelerate the computational speed of solving the mixed integer programming problem by using the branch and cuts strategy [38]. Formulations (8.5)–(8.7) will restrict these continuous variables to be binary value. Formulations (8.9)–(8.11) represent the operational characteristics of individual thermal units, such as ramp rate and minimum on/off time [24–26]. Ramp rate limits restrict the difference of power generations in two adjacent hours. Minimum on/off time represents that the generator has to stay online/offline for several hours before it is turned off/on again. Constraints (8.12)–(8.16) define the bounds for the generations and the reserves for each unit. Only the unit that is offline and has quick start capability within 30 min can provide non-spinning reserve as shown in (8.17). Constraint (8.18) shows that the frequency response reserve from a load resource must be less than its maximum power consumption that is the highest point in its bid curve. The essential scheduling problem in day-ahead market is to balance cleared energy supply with demand, so hourly generation and cleared demand must satisfy the power balance constraint (8.19). Constraint (8.20) represents transmission constraints under normal or contingency condition. Usually, the network security check is separated from unit commitment problem. Only activated transmission constraints will be added into security-constrained unit commitment constraints. Constraints (8.21)–(8.24) denote the traditional regulation up/down and non-spinning reserve requirement constraints in system wide. The penalty prices for unserved reserve are usually high so that reserve will be brought in normal condition.

The minimum requirement of PFR is given in formulation (8.25), which implies that at least some synchronous generating units should be committed to contribute their inertias to the system. The overall FRR requirement $Rfrr_t$ is a function of unit commitment and overall inertias of generating units as shown in the right-hand side of (8.26). The equivalent ratio α_t in (8.27) is a measure scale of FFR performance so that FFR can be equivalently replaced by PFR without sacrificing the performance of PFC capability. Equation (8.27) indicates that the FFR requirement can be determined by the overall FRR need, the equivalency ratio, as well as the amount of PFR. FRR requirement-Inertia and Ratio-Inertia relationship and in (8.26)–(8.27) are a nonlinear function and formulation (8.27) includes bilinear terms which is product of two variables so that the model (8.3)–(8.27) is a mixed integer nonlinear programming which is not effectively solved via commercial solvers, in a reasonable computational time, especially when applied to large-scale power systems. In the next section, we will show how to linearize the FRR requirement-Inertia and the Ratio-Inertia curves and transform bilinear terms by using the linear formulation with big M constraints. In this way, the mixed integer nonlinear programming model can be reformulated into a mixed integer linear or quadratic programming model.

$$\begin{aligned} Max \quad & \sum_{j\epsilon L}\sum_{t\in T}\left[Ce_{j,t}\left(L_{j,t}\right) - Cffr_{j,t}\cdot FFR_{j,t}\right] \\ & - \sum_{i\epsilon G}\sum_{t\in T}\left[STC_{it} + Cf_i\cdot I_{i,t}\cdot LSL_i + Ce_{i,t}(P_{i,t})\right. \\ & \quad + Crup_{i,t}\cdot RUP_{i,t} + Crdn_{i,t}\cdot RDN_{i,t} \\ & \quad\quad \left. + Cnsr_{i,t}\cdot NSR_{i,t} + Cpfr_{i,t}\cdot PFR_{i,t}\right] \\ & - \sum_{t\in T}\left[Nrup_t\cdot RUPN_t + Nrdn_t\cdot RDNN_t\right. \\ & \quad \left. + Nnsr_t\cdot NSRN_t + Npfr_t\cdot PFRN_t + Nfrr_t\cdot FRRN_t\right] \end{aligned} \tag{8.3}$$

s.t.

$$STC_{it} \geq Csu_{i,qc}\cdot\left[Y_{i,t} - \sum_{n=1}^{\min(t,q)} I_{i,t-n}\right], STC_{it}\geq 0 \quad \forall i\in G, t\in T \tag{8.4}$$

$$1 - I_{i,t-1} \geq Y_{i,t} \quad \forall i\in G, t\in T \tag{8.5}$$

$$I_{i,t-1} \geq Z_{i,t} \quad \forall i\in G, t\in T \tag{8.6}$$

$$I_{i,t} - I_{i,t-1} = Y_{i,t} - Z_{i,t} \quad \forall i\in G, t\in T \tag{8.7}$$

$$0 \leq Y_{i,t}, Z_{i,t} \leq 1, I_{i,t}\in\{0,1\} \quad \forall i\in G, t\in T \tag{8.8}$$

$$I_{i,t} \geq \sum_{\tau=\max\left\{1,t-MT_{on,i}+1\right\}}^{t} Y_{i,\tau} \quad \forall i\in G, t\in T \tag{8.9}$$

$$1 - I_{i,t} \geq \sum_{\tau=\max\left\{1,t-MT_{off,i}+1\right\}}^{t} Z_{i,\tau} \quad \forall i\in G, t\in T \tag{8.10}$$

$$-RD_i \leq P_{i,t} - P_{i,t-1} \leq RU_i \quad \forall i\epsilon G, t\in T \tag{8.11}$$

$$P_{i,t} + RUP_{i,t} + PFR_{i,t} \leq HSL_i\cdot I_{i,t} \quad \forall i\epsilon G, t\in T \tag{8.12}$$

$$P_{i,t} - RDN_{i,t} \geq LSL_i\cdot I_{i,t} \quad \forall i\epsilon G, t\in T \tag{8.13}$$

$$0 \leq RUP_{i,t} \leq \overline{RUP}_i\cdot I_{i,t} \quad \forall i\epsilon G, t\in T \tag{8.14}$$

$$0 \leq RDN_{i,t} \leq \overline{RDN}_i\cdot I_{i,t} \quad \forall i\epsilon G, t\in T \tag{8.15}$$

$$0 \leq PFR_{i,t} \leq \overline{PFR}_i\cdot I_{i,t} \quad \forall i\epsilon G, t\in T \tag{8.16}$$

$$0 \leq NSR_{i,t} \leq QSC_i\cdot(1 - I_{i,t}) \quad \forall i\epsilon G, t\in T \tag{8.17}$$

$$0 \leq FFR_{j,t} \leq L_{j,t} \leq MPC_{j,t} \quad \forall j \epsilon D, t \in T \tag{8.18}$$

$$\sum_{i \epsilon G} P_{i,t} = \sum_{j} L_{d,t} \quad \forall t \in T \tag{8.19}$$

$$-\overline{PL}_l \leq \sum\nolimits_{i \in G} SF_{l,i} P_{i,t} + \sum\nolimits_{j \in L} SF_{l,j} L_{j,t} \leq \overline{PL}_l \quad \forall l \in B, t \in T \tag{8.20}$$

$$RUPN_t, RDNN_t, NSRN_t, NPFR_t, FRRN_t \geq 0 \tag{8.21}$$

$$RUPN_t + \sum_{i \epsilon G} RUP_{i,t} \geq Rrup_t \quad \forall t \in T \tag{8.22}$$

$$RDNN_t + \sum_{i \epsilon G} RUP_{i,t} \geq Rrdn_t \quad \forall t \in T \tag{8.23}$$

$$NSRN_t + \sum_{i \epsilon G} NSR_{i,t} \geq Rnsr_t \quad \forall t \in T \tag{8.24}$$

$$NPFR_t + \sum_{i \epsilon G} PFR_{i,t} \geq Rpfr_t \quad \forall t \in T \tag{8.25}$$

$$FRRN_t + FRR_t \geq Rfrr_t \left(\sum_{i \in G} I_{i,t} \cdot H_i \cdot S_i \right) \tag{8.26}$$

$$FRR_t = \sum_{i \epsilon G} PFR_{i,t} + \alpha_t \left(\sum_{i \in G} I_{i,t} \cdot H_i \cdot S_i \right) \times \left(\sum_{j \epsilon D} FFR_{j,t} \right) \tag{8.27}$$

where

i	Index of generating units
j	Index of loads
t	Index of time periods
s	Index of segments
q	Index of segment in stepwise start-up curves
l	Index of transmission branches
P	Cleared energy
L	Cleared demand
I, Y, Z	Binary indicators for unit on/off, start-up and shutdown
δ	Binary variables indicating a segment in a linearized curve is activated
STC	Startup cost of a generating unit
RUP	Regulation up reserve of generating units
RDN	Regulation down reserve of generating units
NSR	Non-spinning reserve of generating units
PFR	Cleared frequency response reserve from primary frequency response of generating units
FFR	Cleared frequency response reserve from fast frequency response of loads
FRR	Total cleared frequency response reserve

$RUPN$	Not served regulation up reserve
$RDNN$	Not served regulation down reserve
$FRRN$	Not served frequency response reserve
$NSRN$	Not served non-spinning reserve
$PFRN$	Not served primary frequency response reserve
Inx	Inertia value of a segment in a linearized curve
β	Variables to replace bilinear terms
α	Equivalent Ratio between FFR and PFR, depending on the total inertias of committed generation units
$Rfrr$	Total requirement of frequency response reserve, depending on the total inertias of committed generation units
Ce	Energy cost/benefit curve based on energy offers/bids
G	Set of generating units
D	Set of demands
T	Set of time periods
B	Set of transmission branches
N	Set of segments
Csu	Step constant in the start-up cost curve of a generating unit
Cf	Minimum energy price
$Nrup$, $Nrdn$	Penalty price for unserved regulation up and down
$Nfrr$, $Nnsr$	Penalty price for unserved frequency responsive reserve and non-spinning reserve
$Npfr$	Penalty price for unserved frequency response reserve and primary frequency response reserve from generating units
$Rrup$, $Rrdn$	Total requirement of regulation up/down reserve
$Rnsr$	Total requirement of frequency responsive reserve and non-spinning reserve
$Rpfr$	Total requirement primary frequency response reserve from generating units
LSL, HSL	Low and high sustainable limits of generating units
MPC	Maximum power consumption for a demand
MT_{on}, MT_{off}	Minimum on/off time of generating units
RU, RD	Maximal ramp up/down limits per hour of generating units
QSC	Quick start capacity of generating units in 30 min
H	Inertial constant
S	Ratepower of generating units
SF	Generator/load shift factor of power network
$\overline{PL}$	Capacity limit of a transmission branch
$\overline{RUP}, \overline{RDN}, \overline{PFR}$	Upper bounds of available regulation up/down reserves and PFR of generating units
$Ratio$	The step value of the Ratio-Inertia curve
$RFRR$.	The point value in vertical scale of the FRR requirement-Inertia curve
In	The inertia value of the linearized curves
M	A larger positive number

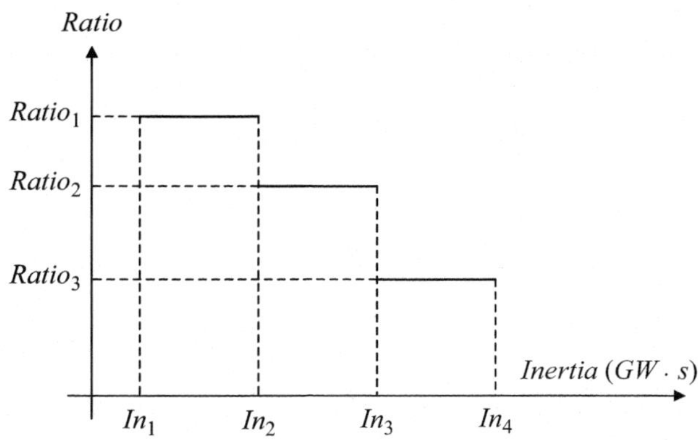

Fig. 8.8 Stepwise ratio-Inertia curve

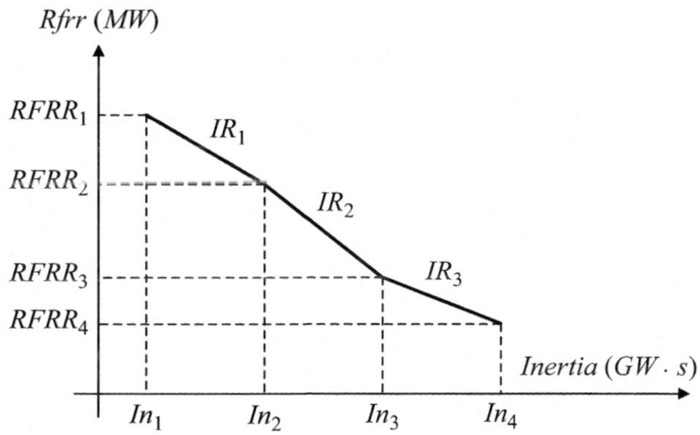

Fig. 8.9 Piecewise linear FRR requirement-Inertia curve

8.5.2 Solution of day-ahead market co-optimization

Discrete points in the FRR requirement-Inertia curve and the Ratio-Inertia curve are obtained from dynamic simulation as shown in Table 8.2. To model it in the market operations, we approximate the Ratio-Inertia and the FRR requirement-Inertia relationships by using stepwise linear curve and piecewise linear curve as displayed in the Figs. 8.8 and 8.9.

In order to represent linearized curves in optimization model, formulations (8.26)–(8.27) are reformulated by (8.28)–(8.32). We introduce additional binary variables to represent activation of a segment in piecewise linear and stepwise curves. The constraint (8.32) denotes that the current segment must be activated if the next segment is activated. Formulation (8.31) defines the upper bound and the lower bound of inertia value of a segment. If the next segment is activated, the current segment must be binding to its maximum value. If the current segment is not activated, the current segment must be equal to zero. Equation (8.30) represents that the total system inertia is equal to the summation of the inertia value of all segments. The FRR requirement for each delivery hour $Rfrr_t$ is equivalent to the right-hand side of (8.29) where IR_s represents the slope of a segment in a piecewise linear curve. The stepwise ratio curve can be represented by the bilinear terms which are the products of the binary variables corresponding every segment and the continuous variables of cleared FFR as shown in (8.28).

$$FRR_t = \sum_{i\epsilon G} PFR_{i,t} + \left[\sum_{s\in N} Ratio_s \cdot (\delta_{s,t} - \delta_{s+1,t})\right] \left(\sum_{j\epsilon D} FFR_{j,t}\right) \quad \forall t \in T \tag{8.28}$$

$$FRRN_t + FRR_t \geq RFRR_1 + \sum_{s\in N} IR_s \cdot Inx_{s,t} \quad \forall t \in T \tag{8.29}$$

$$In_1 + \sum_{s\in N} Inx_{s,t} = \sum_{i\in G} I_{i,t} \cdot H_i \cdot S_i \quad \forall t \in T \tag{8.30}$$

$$(In_{s+1} - In_s) \cdot \delta_{s+1,t} \leq Inx_{s,t} \leq (In_{s+1} - In_s) \cdot \delta_{s,t} \quad \forall s \in N, t \in T \tag{8.31}$$

$$\delta_{s,t} \geq \delta_{s+1,t}, \delta_{s,t} \in \{0,1\} \quad \forall s \in N, t \in T \tag{8.32}$$

However, due to the bilinear terms, the optimization models (8.3)–(8.25) and (8.28)–(8.32) still cannot be solved by commercial solvers effectively. The way to solve this problem is to expand the feasible region by the big M method. We reformulate the bilinear terms in (8.28) by using some linear big M constraints as described in (8.33)–(8.35). If the binary variables $\delta_{s,\,t} - \delta_{s\,+\,1,\,t}$ are equal to one, the constraints (8.34)–(8.35) are relaxed. Otherwise, β_t is binding to the linear term, $Ratio_s \cdot \sum_{j\epsilon D} FFR_{j,t}$. The choice of the big M value is crucial to the computational speed of branch and cuts process to solve the proposed mixed integer linear programming model (8.3)–(8.25) and (8.29)–(8.35). Setting big M too small can lead to infeasible or suboptimal solutions. Therefore, the value of big M will typically have to be rather large in order to exceed the largest activity level. When big M is large, the solver may discover that the feasible region of continues relaxation of the integer programming problem is also large. It can increase upper bound of the mixed integer programming (MIP) (if the objective is to maximize the social welfare) to a significant level and thereby the MIP gap is hardly reduced. In order to make the problem tighter, the value of big M should be as small as possible and exceed the largest activity level. Therefore, we chose $M_{s,t} = Ratio_s \cdot \sum_{j\epsilon D} MPC_{j,t}$ in this paper to make the problem tighter. In practice, since only a portion of load resources may provide FFR, we can reduce the big M to a reasonably small value.

$$FRR_t = \sum_{i\epsilon G} PFR_{i,t} + \beta_t \quad \forall t \in T \tag{8.33}$$

$$\beta_t \geq Ratio_s \cdot \sum_{j \epsilon D} FFR_{j,t} - M_{s,t} \cdot (1 - \delta_{s,t} + \delta_{s+1,t}) \quad \forall s \in N, t \in D \tag{8.34}$$

$$\beta_t \leq Ratio_s \cdot \sum_{j \epsilon D} FFR_{j,t} + M_{s,t} \cdot (1 - \delta_{s,t} + \delta_{s+1,t}) \quad \forall s \in N, t \in D \tag{8.35}$$

8.5.3 Case Studies

The computational study is based on the modified IEEE-118 bus system [23]. The modified IEEE-118 bus power system consists of 54 thermal generating units, 118 buses, and 186 transmission lines. The total installed capacity of 54 thermal generating units is scaled up to 60,000 MW installed thermal generation capacity at ERCOT. The susceptances and thermal rates of transmission branches are also increased in proportion. In this paper, we focus on the modeling of energy and reserve schedule problem with considering high penetration of renewables. Six identical wind farms are added to bus 11, 15, 54, 59, 80, and 90. It is assumed that all wind power offers are \$0.01/MWh. The inertia constant of generating units is given in Fig. 8.10. The maximum capacity for the regulation up reserve, the regulation down reserve, and the PFR reserve of generating units are set as 5%, 5%, and 20% of their high sustainable limit (HSL). The offer prices of the regulation up reserve, the regulation down reserve, the non-spinning reserve, and the PFR reserve are set as 33.3%, 33.3%, 10%, and 20% of the prices of their first energy offer segment. The regulation up/down and non-spinning reserve requirements for different hours in March 2017 can be found in the public market information website of ERCOT [25]. The minimum PFR requirement and FFR requirement are related to the system inertia, which are listed in Table 8.2.

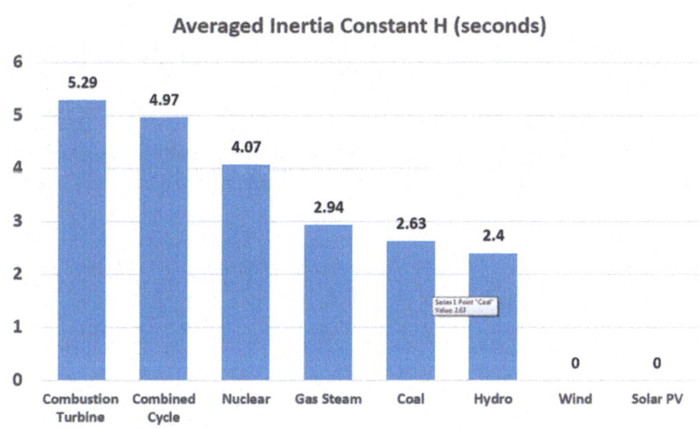

Fig. 8.10 Inertia constant of different generating units

Table 8.3 Bidding parameters and capacities of load resources

Hour 19 (Peak)	Load capacity (MW)	Load bid ($/MWh)	FFR offer ($/MW)
L1	29,406	90	45
L2	8822	35	5
L3	2941	6	3

Table 8.4 Capacity percentages of load resources

Hour	Load capacity percentage (%)	Hour	Load capacity percentage (%)	Hour	Load capacity percentage (%)
1	79	9	76	17	86
2	75	10	79	18	99
3	72	11	81	19	100
4	71	12	83	20	98
5	71	13	84	21	96
6	72	14	84	22	92
7	74	15	83	23	86
8	75	16	82	24	80

We assume that there are three load resources in the day-ahead market. The peak capacity of the load resources is listed in Table 8.3 at hour 19. The load resources are elastic load with bidding prices $90/MWh, $35/MWh, and $6/MWh. The percentage of the capacities of the load resources at different hours is listed in Table 8.4. We assume that all load resources at different hours have the same bidding prices and FFR offer prices.

Three scenarios are simulated, which are described as follows.

Case A: Low penetration of wind generation and low offer prices of FFR. The peak wind-power generation in this base case is 3598 MW at hour 4 so that the wind-power penetration is about 13% at that hour. The hourly maximum wind generation profiles are shown in Table 8.5. The FFR offers of load resources are given in Table 8.3.

Case B: High penetration of wind generation and low offer prices of FFR. The peak wind-power generation in this case is 14,391 MW at hour 4 so that the total wind-power penetration is about 50% at that hour. The hourly maximum wind generation profiles are shown in Table 8.6. The capacities of the load resources and their bid and offer prices are the same as Case A.

Case C: High penetration of wind generation and high offer prices of FFR. The wind-power generation in this case is the same as Case B. The capacities of load resources are the same as Case A. We increase the FFR offer price of L2 to $10/MWh.

All algorithms are implemented in AMPL and solved with CPLEX 12.5. The test environment is AIX server with four 4.024-GHz CPU processors and 64 GB of RAM. The MIP gap is set as 0.6%.

Table 8.5 Total high sustainable limits of low penetration of wind generation

Hour	Wind HSL (MW)	Hour	Wind HSL (MW)	Hour	Wind HSL (MW)	Hour	Wind HSL (MW)
1	3246	7	2230	13	923	19	2239
2	3587	8	2201	14	611	20	2689
3	3530	9	1764	15	1020	21	2903
4	3598	10	1655	16	1056	22	3177
5	3037	11	1243	17	1454	23	3283
6	2631	12	1261	18	1852	24	3319

Table 8.6 Total high sustainable limits of high penetration of wind generation

Hour	Wind HSL (MW)	Hour	Wind HSL (MW)	Hour	Wind HSL (MW)	Hour	Wind HSL (MW)
1	12,985	7	11,016	13	12,870	19	12,534
2	14,347	8	11,526	14	11,868	20	11,184
3	14,121	9	10,584	15	11,370	21	11,610
4	14,391	10	11,370	16	12,036	22	12,709
5	12,148	11	12,078	17	12,792	23	13,132
6	10,524	12	12,228	18	13,548	24	13,275

Table 8.7 Computational time

Wall clock time (s)		Case A	Case B	Case C
With FRR constraints	Original formulation	7.6	84.6	318.9
	Warm start	9.0	7.8	7.5
Without FRR constraints		4.3	3.2	3.2

The social welfare in Case A is \$45,378,657.80 that is lower than that of \$50,218,472.72 in Case B and \$50,177,427.09 in Case C, respectively, because Case B and Case C have more wind energy scheduled in the next day. Furthermore, load resource L2 has the lower FRR offer price in Case B than that in Case C. As a consequence, the social welfare in Case B is slightly higher than Case C. The wall clock time to solve the problem with and without FRR constraints (8.29)–(8.35) is shown in Table 8.7. The computational time of original formulation is longer because of more binary variables and big M constraints. We also propose a warm start process before running the full model of market clearing with FRR constraints. The process is shown in Fig. 8.13. Basically, the market clearing model is executed without FRR constraints. Once the model is solved, the system inertia can be calculated accordingly. Because the total system inertia could not change significantly with FRR constraints, the most of binary variables δ will be fixed except the ones corresponding to the segment of the system inertia resulted from the warm start and its neighboring segments. Therefore, the number of big M constraints will be reduced significantly. With the fixed binary variables and the reduced number of big M constraints, we run the market clearing model with the proposed FRR constraints

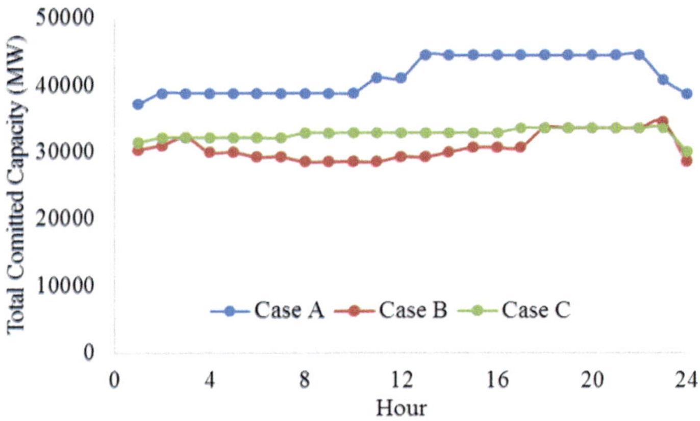

Fig. 8.11 Total committed capacity of thermal generating units

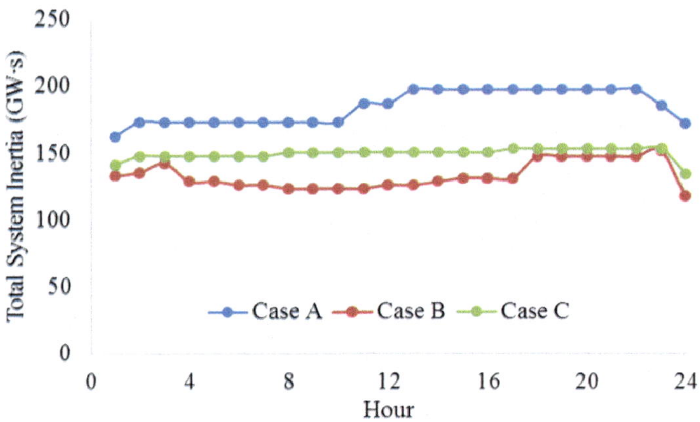

Fig. 8.12 Total system inertia based on unit commitment

to obtain the final solution. The computational time with a warm start is much less than that with the original formulation as listed in Table 8.7.

The energy bid price of the load resource L3 is very low, so L3 is not awarded and the other two loads are awarded in their full bidding capacities in all three cases. The total hourly committed capacities for the three cases are shown in Fig. 8.11. Since the wind generation is low in Case A, the total capacity of committed thermal generating units in Case A is the highest among three cases. Case B and Case C have the same cleared wind generation; however, Case C prefers to commit additional units and thus buys PFR from thermal units rather than obtain FFR from LRs. Case C commits more generating capacity to supply the PFR due to the higher FFR offer price of the load resource L2 in Case C.

The total system inertias based on unit commitment for the three cases are given in Fig. 8.12 in which the trend of the three curves is similar with that in Fig. 8.11. It is

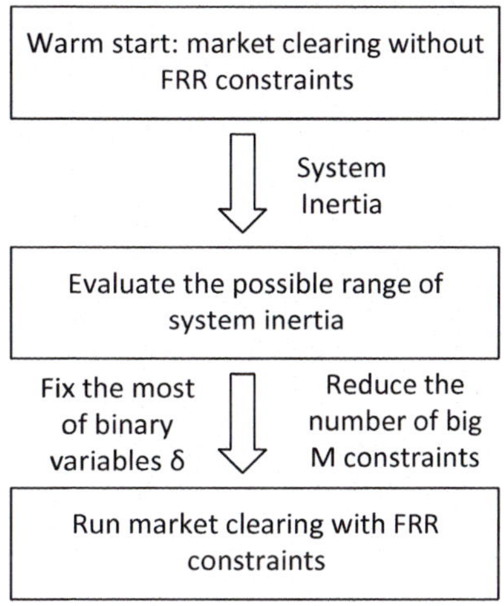

Fig. 8.13 Proposed warm start process

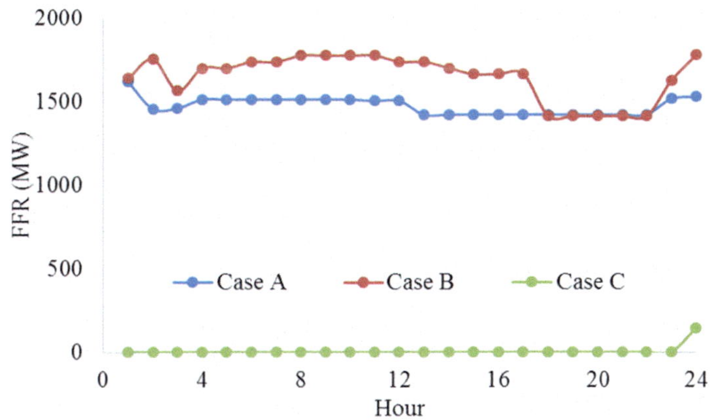

Fig. 8.14 Cleared FFR in day-ahead market

noted that the higher system inertia can result in a lower FRR requirement and a lower PFR/FFR ratio. As a result, the FRR requirement under high wind penetration condition is higher than that under low wind penetration condition. Therefore, more FFR are awarded in Case B than Case A while the cleared PFR are almost the same in the two cases as shown in Figs. 8.14 and 8.15. In Case C, the thermal generating units provide all FRR except the hour 24. The load resource L2 provides additional FRR at hour 24 because all PFR are awarded and committing additional thermal generating units is not economical.

The cleared prices for PFR and FFR are shown in Figs. 8.16 and 8.17. The PFR price is equal to the sum of the dual variables of the constraints (8.25) and (8.26). If the thermal generating units are the marginal resources to supply FRR, the PFR price will reflect the PFR offer of those marginal resources such as hour 1–24 in Case C. If

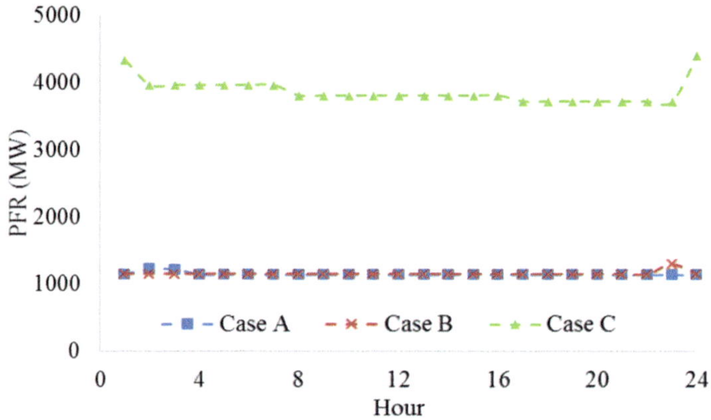

Fig. 8.15 Cleared PFR in day-ahead market

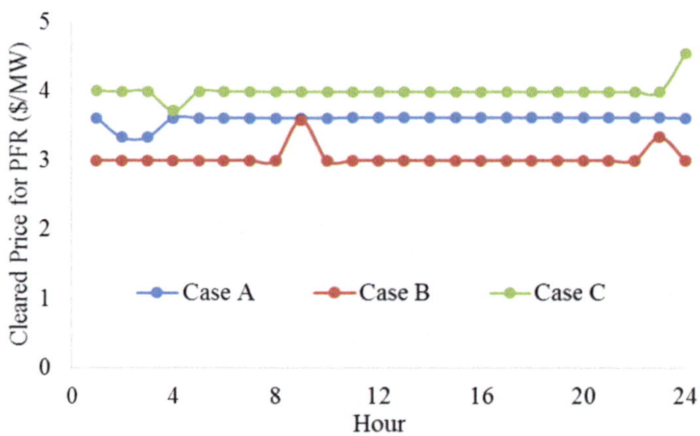

Fig. 8.16 Cleared price for PFR in day-ahead market

the load resources are marginal resources to supply FRR, both PFR offer and FFR offer will impact the PFR cleared price because there is the minimum PFR requirement constraint (8.25). It is possible that the minimum PFR requirement constraint is binding and the dual variable of (8.25) is not zero, such as the most hours in Case A and Case B. The cleared FFR price accounts for the marginal offer of FRR and the equivalency ratio between the FFR and PFR. During hour 1 to hour 23 in Case C, there is no awarded FFR since the cleared FFR price is lower than the FFR offer. In Cases A and B, the FFR offer will determine the cleared price of FFR because load resource L2 is the marginal resource to provide FRR.

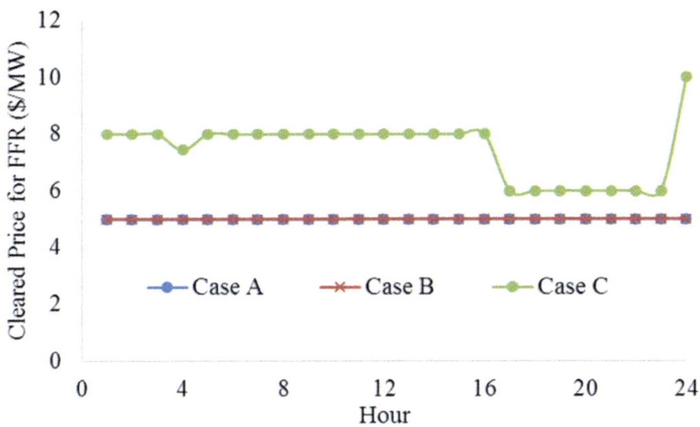

Fig. 8.17 Cleared price for FFR in day-ahead market

8.6 Co-optimization of Energy and FRR in Real-time

The previous section has outlined the main design principles of a day-ahead (DA) PFC market, and this problem is essentially formulated as a unit commitment problem. In this section, a new real-time (RT) PFC market mechanism for accommodating FRR offers from both load and generation resources is proposed, which is solved as security-constrained economic dispatch (SCED).

The unique features of the new market are as follows. First, from the DA unit commitment results, the system inertia for the operational hour can be determined, and in turn, the FRR requirement for the corresponding hour is estimated as a prior. As a result, system inertia conditions are no longer part of the market constraints or the decision variable in real-time economic dispatch because they are replaced by the predetermined hourly FRR needs. Second, the new market mechanism enables the PFR and FFR to compete in a single market. When clearing FRR offers based on the marginal cost for providing one unit of the FRR service, the interdependency and equivalency between PFR and FFR for each inertia condition is considered. Because loads and generators are now considered as substitutable FRR resources that are cleared at a single market, the new market mechanism enables energy and FRR to be co-optimized in the real time. Third, the flexible load resources can adjust their energy bid and FFR offers easily and more adaptively to the real-time operational conditions. The market outcome resulting from different bidding strategy of load resources should be anticipated to ensure liquidity of the market and an efficient way exists for load resources to hedge against the risk of participating in the both RT energy and FRR market.

To simplify the problem formulation, the following assumptions are made:

1. The clearance of DA market is known.

2. Remaining energy needs and AS products are cleared by economic dispatch at the RT market based on the unit commitment results obtained from the DA clearing process.
3. The network model is neglected. At the ERCOT ancillary service market, the price and quantities are cleared without a network model with an underlying assumption that the generation units or load resources providing AS are not behind a constrained path. The occurrence of congestions on the transmission lines have greatly reduced in the ERCOT area since the project of the Competitive Renewable Energy Zone was in operation in 2014. Note that the proposed formulation can be readily applied to include a network model to account for the locational difference for LMP for the AS service.
4. The energy offers and AS bids are constant, i.e., the load and generators will submit a single set of energy and AS bids.

Thus, a hypothetical, single-period economic dispatch model for co-optimization of energy, PFR, and FFR can be formulated, which is described in the next subsection. While AS products considered here only include PFR and FFR, other AS products such as regulation and non-spinning reserve services can be readily included in the problem formulation.

8.6.1 Problem Formulation

The objective function and constraints of the co-optimization problem are represented as follows:

$$\max\left\{\sum_{j=1}^{Nd} b_j(d_j) - \sum_{i=1}^{Ng} f_i(p_i) - \sum_{i=1}^{Ng} f_i^{PFR}\left(p_i^{PFR}\right) - \sum_{j=1}^{Nd} f_j^{FFR}\left(p_j^{PFR}\right)\right\} \tag{8.36}$$

subject to the following constraints:

Power Balance Constraint

$$\sum_{i=1}^{Ng} p_i = \sum_{j=1}^{Nd} d_{\mathrm{j}}\ (\lambda) \tag{8.37}$$

Frequency Response Reserve Constraint

$$\sum_{i=1}^{Ng} p_i^{PFR} + m \times \sum_{j=1}^{Nd} p_j^{FFR} \geq FRR_{min}\ (\beta) \tag{8.38}$$

Primary Frequency Response Constraint

$$\sum_{i=1}^{Ng} p_i^{PFR} \geq PFR_{\min}\ (\alpha) \tag{8.39}$$

Generation Capacity Constraint

$$\begin{pmatrix} \underline{\rho_{i1}} \\ \underline{\rho_{i2}} \\ \underline{\rho_{i3}} \end{pmatrix} \begin{bmatrix} LSL_i - P_i^{DA} \\ \underline{PFR_i} \\ LSL_i - P_i^{DA} \end{bmatrix} \leq M \times \begin{bmatrix} p_i \\ p_{i^{PFR}} \end{bmatrix} \leq \begin{bmatrix} HSL_i - P_i^{DA} \\ \overline{PFR_i} \\ HSL_i - P_i^{DA} \end{bmatrix} \begin{pmatrix} \overline{\rho_{i1}} \\ \overline{\rho_{i2}} \\ \overline{\rho_{i3}} \end{pmatrix} \tag{8.40}$$

$$M = \begin{bmatrix} 1 & 0 \\ 0 & 1 \\ 1 & 1 \end{bmatrix}, M_1 = \begin{bmatrix} 1 \\ 0 \\ 1 \end{bmatrix}, M_2 = \begin{bmatrix} 0 \\ 1 \\ 1 \end{bmatrix}, [\underline{\rho_i}] = \begin{pmatrix} \underline{\rho_{i1}} \\ \underline{\rho_{i2}} \\ \underline{\rho_{i3}} \end{pmatrix}, [\overline{\rho_i}] = \begin{pmatrix} \overline{\rho_{i1}} \\ \overline{\rho_{i2}} \\ \overline{\rho_{i3}} \end{pmatrix}$$

Because P_i^{DA} is a quantity cleared at the DA market by unit commitment, it is not a decision variable in the model.

Committed Load Limit Constraint

$$\left(\underline{\theta_j}\right) \quad 0 \leq d_j \leq D_j \quad \left(\overline{\theta_j}\right) \tag{8.41}$$

Fast Frequency Response Limit Constraint

$$\left(\underline{\eta_j}\right) \quad 0 \leq FFR_j \leq d_j + D_j^{DA} \quad \overline{\left(\eta_j\right)} \tag{8.42}$$

Similarly, because D_j^{DA} is cleared at the DA market, it is not a decision variable in the RT model. However, the amount of the FFR provided by a load resource cannot exceed the total of its cleared amount at both the DA and RT market because the FFR is deployed by shedding a portion of the committed load.

where

i	Index for all units
j	Index for all units
N_g	Number of units
N_d	Number of loads
P_{PFR}	Total amount of PFR needed (MW)
P_{FFR}	Total amount of FFR needed (MW)
$FRR_{\min}$	Minimum amount of FRR required (MW)
$PFR_{\min}$	Minimum amount of PFR required (MW)
m	Equivalency ratio between PFR and FFR
p_i	Energy offer award for unit i in real-time market (MW)
d_j	Energy bid award for load j in real-time market (MW)
$P^{DA}{}_i$	Energy offer award for unit i in day-ahead market (MW)
$D^{DA}{}_j$	Energy bid award for load j in day-ahead market (MW)
$p^{PFR}{}_i$	PFR award for unit i (MW)

p^{FFR}_j	FFR award load j (MW)
$f_i(p_i)$	Energy offer for unit i in real-time market (\$/MWh)
$b_j(d_j)$	Energy bid for load j in real-time market (\$/MWh)
$f_i^{PFR}(p_i^{PFR})$	PFR offer for unit i in real-time market (\$/MW)
$f_j^{FFR}(p_j^{FFR})$	FFR offer for load j in real-time market (\$/MW)
HSL_i	High sustainable limit for unit i (MW)
LSL_i	Low sustainable limit for unit i (MW)
D_j	Total energy bid for load j in real-time market (MW)
$\overline{PFR_i}$	High limit of PFR for unit i (MW)
$\underline{PFR_i}$	Low limit of FFR for unit i (MW)
λ	Shadow price of power balance constraint
β	Shadow price of FRR requirement constraint
α	Shadow price of PFR requirement constraint
$\underline{\rho_i}$	Shadow price of LSL constraint for unit i
$\overline{\rho_i}$	Shadow price of HSL constraint for unit i
$\underline{\theta_j}$	Shadow price of LSL constraint for load j
$\overline{\theta_j}$	Shadow price of HSL constraint for load j
$\underline{\eta_j}$	Shadow price of FFR lower constraint for load j
$\overline{\eta_j}$	Shadow price of FFR upper constraint for load j

8.6.2 Pricing Strategy

The energy price is set as λ, which is the shadow price calculated from the power balance constraint (8.37). The shadow price for PFR is β for meeting the minimum FRR requirement (8.38) and α for meeting the minimum PFR requirements (8.39). For example, if the cleared amount of PFR exceeds the minimum PFC requirement, minimum PFR constraint (8.39) will not be a binding constraint, so $\alpha = 0$. Otherwise, α is the shadow price for procuring the minimum amount of PFR. The final PFR shadow price is $\alpha + \beta$. The FFR price is $m{\cdot}\beta$ to account for the equivalency ratio between the FFR and PFR. The pricing scheme is summarized in Table 8.8 and its optimality is achieved for ensuring the profit maximization for both generation and load resources when participating both the energy and the newly proposed PFC market.

Table 8.8 Shadow price calculation for each FRR resource

Product	Price	Constraints
Energy	λ	Power balance equation
PFR	$\beta + \alpha$	Minimum FRR and PFR requirement
FFR	$m{\cdot}\beta$	Minimum FRR requirement

Table 8.9 Bidding parameters and capacities of generation resources

Generator	Capacity (MW)	DA Cleared Capacity (MW)	RT Energy Offers ($/MWh)	PFR Capacity (MW)	PFR offers ($/MW)
G1	[0,5000]	0	7000	[0,1000]	20
G2	[0,15,000]	5600	50	[0,3000]	15
G3	[0,20,000]	14,000	20	[0,4000]	10
G4	[0,10,000]	7000	10	–	–
G5 (Wind)	[0,2000]	1400	0.01	–	–

Table 8.10 Bidding parameters and capacities of load resources

Load	Total Load (MW)	DA Cleared Load (MW)	RT Energy Bids ($/MWh)	FFR Capacity (MW)	FFR offers ($/MW)
L1	26,000	18,200	9000	–	–
L2	8000	5600	80	[0,8000]	30
L3	6000	4200	60	[0,6000]	**18/20/26**
L4	400	0	30	[0,400]	4
L5	200	0	25	[0,200]	3

8.6.3 *FFR's Market Behaviors*

In this section, we use a simplified system with five generators (G1–G5) and five loads (L1–L5) to demonstrate basic operation principles of the newly proposed PFC market.

8.6.3.1 Base Case

Because a load provides FFR by reducing its energy consumption, only when the load committed at the DA or RT energy market can provide FFR to the PFC market, as bonded by constraints (8.41) and (8.42). Therefore, the base case study will illustrate three pricing conditions: a load already committed at the DA energy market offers FFR at a much cheaper price than that of the generator provided PFC; a load already committed at the DA energy market offers FFR that is competitive to the generator-provided PFC, and a load that would not have been able to commit at both the DA and RT energy market is committed to the RT energy market because of offering competitive PFR service.

The parameters of the base case are selected so that the net-load is between 35 and 40 GW, $FRR_{\min} = 3040$ MW, $m = 1.25$, and $PFR_{\min} = 1143$ MW. The generator and load capacity cleared at the DA market and their RT offers are listed in Tables 8.9 and 8.10. G4, G5, and L1 will not bid into the PFC market while L4 and L5 are outbid at the DA energy market. To simplify the illustration, we only alter the FFR

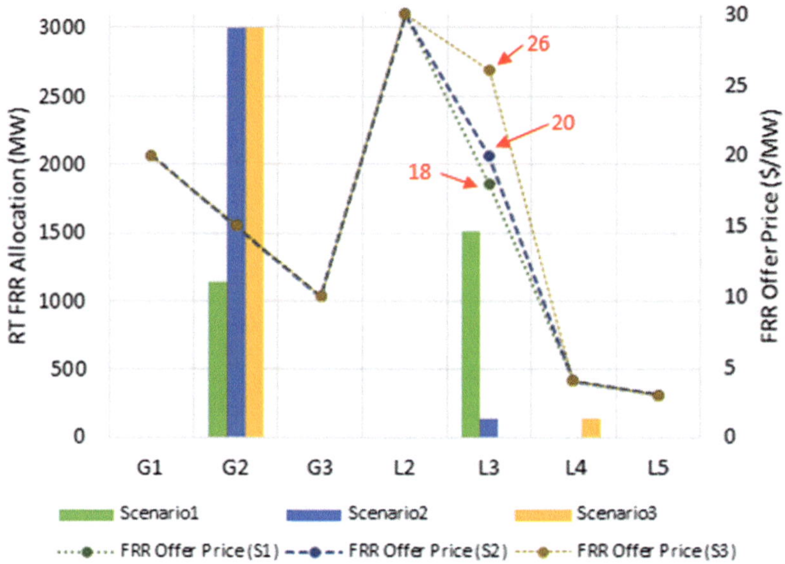

Fig. 8.18 FRR allocation with L3 FFR offer changes

offers for L3. The results are shown in Table 8.11. Note that the energy and PFC markets are cleared by the pricing scheme listed in Table 8.11. Specifically, because 1 MW of FFR can replace m MW of PFR (1.25 MW in this case), FFR providers are rewarded m more than PFR providers. Therefore, the FFR clearing price is 1.25 times higher than the PFR clearing price.

Figure 8.18 illustrates how the change of the L3 FFR offer affects the FRR allocation among different resources.

In Scenario 1, if the FFR offer ($18/MW) from L3 is very low, L3 can provide all the FRR service needed ($FRR_{\min} = 3040$ MW). However, because of the minimum PFR requirement ($PFR_{\min} = 1143$ MW), G2 is awarded 1143 MW. Because (8.39) is binding, $\alpha = 0.4$. In Scenario 2, if the FFR offer ($20/MW) from L3 is competitive with the PFR offers provided by generators, G2 will be awarded 3000 MW. Again, L3 will be the marginal resources. Because (8.39) is no longer binding, $\alpha = 0$. In Scenario 3, if the FFR offer ($26/MW) from L3 is high, G2 will again be awarded 3000 MW but L3 is outbid by L4. L4 was outbid in the first two scenarios by L3 because its energy bid ($30/MW) is lower than the energy clearing price ($50/MW), so it didn't qualify for providing FRR. When FRR is cleared at $24/MW, L4 will be accepted at the energy market because the extra energy payment (i.e., 50–30 = 20 $/MW) can be subsided by its income (i.e., $24/MW) at the PFC market. Thus, a load offer that would have been outbid at the RT energy market will have a chance to be accepted by providing competitive FFR service to the PFC market. This is one of the main merits of the proposed new PFC market pricing scheme.

Table 8.11 Energy and FRR allocation

	Scenario1	Scenario2	Scenario3
Energy/PFR (G1)	0/0	0/0	0/0
Energy/PFR (G2)	2400/**1143**	2400/**3000**	2432/**3000**
Energy/PFR (G3)	6000/0	6000/0	6000/0
Energy/PFR (G4)	3000/–	3000/–	3000/–
Energy/PFR (G5)	600/–	600/–	600/–
Energy/FFR (L1)	7800/–	7800/–	7800/–
Energy/FFR (L2)	2400/0	2400/0	2400/0
Energy/FFR (L3)	1800/**1517.6**	1800/**32**	1800/0
Energy/FFR (L4)	0/0	0/0	32/**32**
Energy/FFR (L5)	0/0	0/0	0/0
Energy Clearing Price	$\lambda = 50$($/MWh)	$\lambda = 50$ ($/MWh)	$\lambda = 50$($/MWh)
PFR Clearing Price	$\alpha + \beta = 15$($/MW)	$\alpha + \beta = 16$($/MW)	$\alpha + \beta = 19.2$($/MW)
FFR Clearing Price	$m\beta = 18$($/MW)	$m\beta = 20$($/MW)	$m\beta = 24$($/MW)
λ, α, β	50,0.6,14.4	50,0,16	50,0,19.2

8.6.3.2 Mitigation of Price Spikes at the PFC Market

The main advantage of letting loads offer FRR is to introduce competition into the PFC market and mitigate price spikes at the RT energy and PFC markets. Tables 8.12 and 8.13 summarize the bidding parameters for the price mitigation scenarios. Let L1-L3 be must-serve loads so they will not participate into the PFC market. The results of the price mitigation case studies are shown in Table 8.14.

Case 1: Energy and PFC Market Prices Without FFR Offers
We do not allow L4 to bid into the PFC market and FRR can only be provided by generators. When cheap generation resources are not sufficient at either market, both the energy and FRR prices will spike. This is because the two markets are co-optimized and so opportunity costs will incur. As shown by the third column in Table 8.14 G1, the most expensive unit, sets energy price such that $\lambda = 120\$/MW$. G2 has reached its generation limit by providing 8160 MW of energy and 840 MW of PFR. Because (8.40) is binding for G2, there is an opportunity cost of $\overline{\rho_{23}} = \lambda - \nabla f_2(p_2) = 120 - 50 = 70\$/MW$, which represents the cost of compensating G2 for supplying PFR instead of energy. As G2 clears the PFC market, the PFC market clearing price is calculated as $\alpha + \beta = \nabla f_2^{PFR}\left(p_2^{PFR}\right) + \overline{\rho_{23}} = 15 + 70 - 85\$/MW$. In this case, both L4 and L5 are outbid at the RT energy market.

Case 2: Energy and PFC Market Prices With FFR Offers
To mitigate the PFC market price spike, we allow L4 to provide FFR and bid into the PFC market. By varying L4 FFR offering price from \$1/MW to \$30/MW, we recalculate the energy and the PFC market prices and plot the results in Fig. 8.19. As shown in the figure, introducing FFR will considerably reduce the prices in both the energy and the PFC market. When the offering price is \$1/MW, L4 will be committed to both the energy and the PFC markets as shown in Table 8.14. The

Table 8.12 Bidding parameters and capacities of generation resources

	Capacity (MW)	DA cleared capacity (MW)	RT energy offers ($/MWh)	PFR capacity (MW)	PFR offers ($/MW)
G1	[0,11,000]	28	120	[0,2200]	20
G2	[0,9000]	5712	50	[0,1800]	15
G3	[0,20,000]	14,000	20	[0,4000]	10
G4	[0,10,000]	7000	10	–	–
G5 (Wind)	[0,2000]	1400	0.01	–	–

Table 8.13 Bidding parameters and capacities of load resources

	Total Load (MW)	Cleared Load at DA (MW)	Energy Bids at RT ($/MWh)	FFR Capacity (MW)	FFR Offers ($/MW)
L1(must serve)	26,200	18,340	9000	–	–
L2(must serve)	8000	5600	8000	–	–
L3(must serve)	6000	4200	8000	–	–
L4	400	0	30	[0,400]	**variable**
L5	200	0	25	–	–

Table 8.14 Energy and FRR allocation (Price Spike)

	L4 FFR Offer $1/MW	L4 FFR Offer $8/MW	No FFR or L4 FFR Offers >$17/MW
Energy/PFR (G1)	0/**2200**	0/**2200**	12/**2200**
Energy/PFR (G2)	2508/**780**	2508/**780**	2448/**840**
Energy/PFR (G3)	6000/0	6000/0	6000/0
Energy/PFR (G4)	3000/–	3000/–	3000/–
Energy/PFR (G5)	600/–	600/–	600/–
Energy/FFR (L1)	7860/–	7860/–	7860/–
Energy/FFR (L2)	2400/–	2400/–	2400/–
Energy/FFR (L3)	1800/–	1800/–	1800/–
Energy/FFR (L4)	48/**48**	48/**48**	0/0
Energy/FFR (L5)	0/0	0/0	0/0
Energy Clearing Price	$\lambda = 59$($/MWh)	$\lambda = 87$($/MWh)	$\lambda = 120$($/MWh)
PFR Clearing Price	$\alpha + \beta = 24$($/MW)	$\alpha + \beta = 52$($/MW)	$\alpha + \beta = 85$($/MW)
FFR Clearing Price	$m\beta = 30$($/MW)	$m\beta = 65$($/MW)	$m\beta = 106.25$($/MW)
λ, α, β	59,0,24	87,0,16	120,0,85

$m = 1.25$ $FRR_{min} = 3040$ MW $PFR_{\min} = 1143$ MW

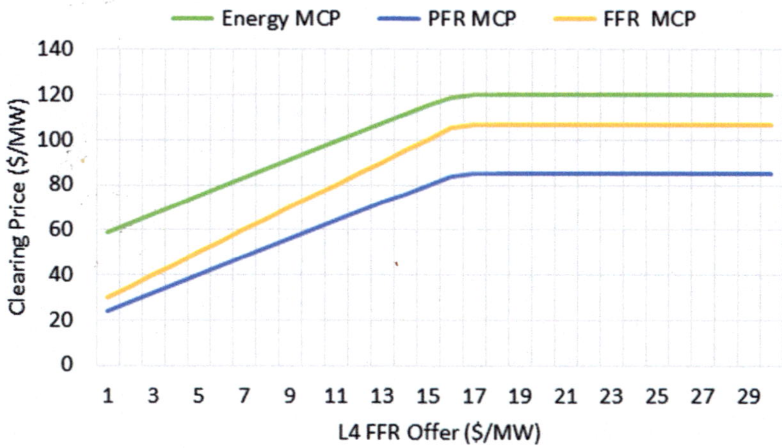

Fig. 8.19 Energy and FRR prices when L4 FFR offer raises from \$1/MW to \$30/MW

prices at the energy and the PFC markets will drop from \$120/MW to \$60/MW and \$85/MW to \$22/MW, respectively. However, when L4 increases its bid from \$1/MW to \$17/MW, the resulting price mitigation effect is diminishing. This is because when the cost for L4 to provide FFR increases to a certain point, the revenue it receives for providing FFR can no longer offset the additional payment made to purchase energy. Thus, when the FFR offer reaches \$17/MW, L4 will be outbid at both the energy and the PFC markets. As a result, the FRR will again be fully provided by G1 and G2. The FRR allocations when the L4 FFR offering price is \$1/MW, \$8/MW, and \$17/MW (and above) are shown in Table 8.14. The results further prove that providing FFR increases the chance for a load to be served at the RT energy market.

8.7 Conclusions

Increasing growth of renewable resources could lead to a declined system inertia, which poses a threat to the grid security and reliability. Among various solutions proposed, to maintain sufficient PFC capability is crucial under high penetration of renewable energy condition. However, to allocate such a PFC capacity from the generation side will come at a high cost especially when the online reserve is scarce. Alternatively, load resources can provide a comparable, or even better, primary frequency response. This will benefit a future power grid both reliably and economically by lifting the burden partially from the generation side. The operation paradigm for the future grid will be also shifted if an FRR market is established. This will incentivize these resources which can contribute to the system reliability when needed and reward them based on their performance. In the long term, it will attract more participants to compete and the market efficiency will be eventually improved.

Based on a case study using ERCOT data and model, the advantages of creating an FRR market are found to be tremendous. Numerical simulations show that PFR and FFR can be substituted with an equivalency ratio, which is largely impacted by system inertia. Since this dependency is nonlinear, it can be approximated as stepwise ratio-inertia curve and piecewise linear FRR requirement-inertia curve. After this simplification, the co-optimization model in a DA market can be solved by a linear reformulation technique with big M. These case studies show the effectiveness of the proposed solution and the correctness of the cleared reserve and prices.

In an RT market, after the equivalency between the fast frequency response provided by loads or other resources and the primary frequency response provided by generators for typical system inertia conditions is quantified, the pricing scheme of a new PFC market for accommodating both the load and generator FRR resources can be derived. The pricing scheme is able to reflect the performance equivalency between different FRR resources and ensure that their profits are maximized in both the energy and the ancillary service markets. Simulation results demonstrate that by allowing the load to provide fast frequency response, price spikes can be mitigated. In addition, by participating in the PFC market and receiving a reward for the AS service, loads previously outbid at the energy market can be served.

These discussions clearly lay out a path of operating a competitive PFC market to integrate large-scale renewable resources. While this work is primarily focused on frequency-responsive LRs, the fundamental principle can be extended to other resources which are also able to respond to the large frequency deviations, such as residential load, energy storages, and synthetic inertia from wind turbines. Active research work is also being pursued to exploit the potential of aggregating a large number of residential load to provide primary frequency control, and the results are very encouraging. As more efficient and less costly resources can respond to the large frequency deviations, the declining inertia driven by the high penetration of renewable resources will not become a concern so that more renewables can be fully integrated to the grid.

References

1. Kundur, P. (1994). *Power system stability and control*. New York: McGraw-Hill.
2. Bevrani, H. (2009). *Robust power system frequency control*. New York: Springer.
3. Ela, E., Milligan, M., Kirby, B., Tuohy, A., & Brooks, D. (2012). *Alternative approaches for incentivizing the frequency responsive reserve ancillary service*. NREL.
4. Eto, J. H. (2011). Use of frequency response metrics to assess the planning and operating requirements for reliable integration of variable renewable generation. *Lawrence Berkeley National Laboratory*.
5. NERC Frequency Response Initiative. (2010).

6. Sharma, S., Huang, S.-H. & Sarma, N. D. R. (2011). System inertial frequency response estimation and impact of renewable resources in ERCOT interconnection. *2011 IEEE Power and Energy Society General Meeting* (pp. 1–6), San Diego, CA.
7. Doherty, R., Mullane, A., Nolan, G. L., Burke, D. J., Bryson, A., & O'Malley, M. (2010). An assessment of the impact of wind generation on system frequency control. *IEEE Transactions on Power Systems, 25*(1), 452–460.
8. O'Sullivan, J., et al. (2014). Studying the maximum instantaneous non-synchronous generation in an island system—Frequency stability challenges in Ireland. *IEEE Transactions on Power Systems, 29*(6), 2943–2951.
9. Huang, S.-H., Dumas, J., González-Pérez, C., & Lee, W.-J. (2009). Grid security through load reduction in the ERCOT market. *IEEE Transactions On Industry Applications, 45*(2), 555–559.
10. Morren, J., Pierik, J., & De Haan, S. W. H. (2006). Inertial response of variable speed wind turbines. *Electric Power Systems Research, 76*(11), 980–987.
11. Morren, J., et al. (2006). Wind turbines emulating inertia and supporting primary frequency control. *IEEE Transactions on Power Systems, 21*(1), 433–434.
12. Delille, U., Francois, B., & Malarange, G. (2012). Dynamic frequency control support by energy storage to reduce the impact of wind and solar generation on isolated power system's inertia. *IEEE Transactions on Sustainable Energy, 3*(4), 931–939.
13. Ma, O., Alkadi, N., Cappers, P., Denholm, P., Dudley, J., Goli, S., Hummon, M., Kiliccote, S., MacDonald, J., Matson, N., Olsen, D., Rose, C., Sohn, M. D., Starke, M., Kirby, B., & O'Malley, M. (2013). Demand response for ancillary services. *IEEE Transactions on Smart Grid, 4*(4), 1988–1995.
14. Molina-García, A., Bouffard, F., & Kirschen, D. S. (2011). Decentralized demand-side contribution to primary frequency control. *IEEE Transactions on Power Systems, 26*(1), 411–419.
15. *Autonomous demand response for primary frequency regulation*. Pacific Northwest National Laboratory, 2012.
16. Ela, E., Gevorgian, V., Tuohy, A., Kirby, B., Milligan, M., & O'Malley, M. (2014). Market designs for the primary frequency response ancillary service—Part I: motivation and design. *IEEE Transactions on Power Systems, 29*(1), 421–431.
17. Ela, E., Gevorgian, V., Tuohy, A., Kirby, B., Milligan, M., & O'Malley, M. (2014). Market designs for the primary frequency response ancillary service—part II: case studies. *IEEE Transactions on Power Systems, 29*(1), 432–440.
18. O'Sullivan, J. W., & O'Malley, M. J. (1996). Economic dispatch of a small utility with a frequency based reserve policy. *IEEE Transactions on Power Systems, 11*(3), 1648–1653.
19. O'Sullivan, J. W., & O'Malley, M. J. (1999). A new methodology for the provision of reserve in an isolated power system. *IEEE Transactions on Power Systems, 14*(2), 519–524.
20. Papadogiannis, K. A., & Hatziargyriou, N. D. (2004). Optimal allocation of primary reserve services in energy markets. *IEEE Transactions on Power Systems, 19*(1), 652–659.
21. Galiana, F. D., Bouffard, F., Arroyo, J. M., & Restrepo, J. F. (2005). Scheduling and pricing of coupled energy and primary, secondary, and tertiary reserves. *Proceedings of the IEEE, 93*(11), 1970–1983.
22. Doherty, R., Lalor, G., & O'Malley, M. (2005). Frequency control in competitive electricity market dispatch. *IEEE Transactions on Power Systems, 20*(3), 1588–1596.
23. Restrepo, J. F., & Galiana, F. D. (2005). Unit commitment with primary frequency regulation constraints. *IEEE Transactions on Power Systems, 20*(4), 1836–1842.
24. *NERC Reliability Standard BAL-003 frequency response and frequency bias setting, NERC.*
25. ERCOT FAST Consolidated Working Document 5–16-14.
26. Third-Party Provision of Primary Frequency Response Service, FERC, November 20, 2015.
27. Pengwei, D., Makarov, Y. V., Pai, M. A., & McManus, B. (2014). Calculating individual resources variability and uncertainty factors based on their contributions to the overall system balancing needs. *IEEE Transactions on Sustainable Energy, 5*(1), 323–331.
28. Du, P., & Matevosyan, J. Forecast system inertia condition and its impact to integrate more renewables. *IEEE Transactions on Smart Grid, 2017.*

29. Du, P., Hui, H., & Lu, N. (2016). Procurement of regulation services for a grid with high-penetration wind generation resources: a case study of ERCOT. *IET Generation, Transmission & Distribution, 10*(16), 4085–4093.
30. Du, P., & Makarov, Y. (2014). Using disturbance data to monitor primary frequency response for power system interconnections. *IEEE Transactions on Power Systems, 29*(3), 1431–1432.
31. Li, W., Du, P., & Lu, N. (2017). *Design of a new primary frequency control market for hosting frequency response reserve offers from both generators and loads*. IEEE Transactions on Smart Grid. doi: https://doi.org/10.1109/TSG.2017.2674518
32. Liu, C., Botterud, A., Zhou, Z., & Du, P. (2017). Fuzzy energy and reserve co-optimization with high penetration of renewable energy. *IEEE Transactions on Sustainable Energy, 8*(2), 782–191.
33. Wood, A. J., & Wollenberg, B. F. (1996). *Power generation, operation and control* (2nd ed.). New York, NY: Wiley.
34. Shahidehpour, M., Yamin, H., & Li, Z. (2002). *Market Operations in Electric Power Systems*. New York, NY: Wiley.
35. Liu, C., Shahidehpour, M., Li, Z., & Fotuhi-Firuzabad, M. (2009). Component & mode models for short-term scheduling of combined-cycle units. *IEEE Transactions on Power Apparatus and Systems, 24*, 976–990.
36. Lee, C., Liu, C., Mehrotra, S., & Shahidehpour, M. (2014). Modeling transmission line constraints in two-stage robust unit commitment problem. *IEEE Transactions on Power Systems, 29*(3), 1221–1231.
37. Tan, Y. T., & Kirschen, D. 1 S. (2006). Co-optimization of energy and reserve in electricity markets with demand-side participation in reserve services. *Power Systems Conference and Exposition, 2006*. PSCE'06. 2006 IEEE PES. IEEE, 2006.
38. Rajan, D., & Takriti, S. (2005). *Minimum up/down polytopes of the unit commitment problem with start-up costs*. IBM Research Report.
39. Bhana, R., & Overbye, T. J. (2016). The commitment of interruptible load to ensure adequate system primary frequency response. *IEEE Transactions on Power Systems, 31*(3), 2055–2063.

Chapter 9
Optimal Response of Residential House Load

9.1 Introduction

Traditionally, a power system is only scheduled for operation on generation resources because the majority of power system loads are neither controllable nor measurable at the required time resolution. In addition, the non-differential retail electricity price provides little incentive for consumers to schedule their energy consumption.

Millions of smart meters, sensors, and automatic control devices were deployed in electric power distribution grids down to residential and commercial buildings via two-way communication networks. Retail pricing schemes, such as real-time pricing (RTP), time-of-use (TOU), and critical peak pricing (CPP), are being designed by utilities to provide incentives for consumers to actively participate in demand response (DR) programs. Furthermore, it is possible to control and monitor the on/off, cycling, or mode switching of appliances wirelessly with a home energy management system by developing smart appliances and deploying home area networks. For example, the thermostat settings of a heating, ventilating, and air conditioning (HVAC) system can be programmed remotely via a ZigBee network; a smart dryer can be switched from the "normal" mode to "energy saving" modes to reduce its energy consumption.

The smart grid technologies are enablers to schedule loads at the consumer level to save energy, reduce cost, and help grid operation; however, it is impractical to request a consumer, who is neither an economist nor an experienced grid operator, to create an optimal schedule from the many possibilities. Therefore, an automatic load scheduling method that requires little consumer attention to set up and maintain and allows consumers to compare costs/benefits with different load schedules, is needed to gain consumer acceptance on consumer-level DR programs. It is also needed to motivate consumers' active participation in shaping their energy consumption for the benefit of themselves as well as for the efficient, reliable, and economical operation of the power grid.

P. Du et al., *Demand Response in Smart Grids*,
https://doi.org/10.1007/978-3-030-19769-8_9

The past works have explored automatic ways to determine the energy schedule for residential house customers while considering the users' preference and other constraints. One of these schemes is introduced here, which is a novel appliance commitment algorithm that schedules thermostatically controlled appliances (TCAs) based on price and consumption forecasts.

The appliance commitment, similar to the unit commitment problem [1], finds an optimal schedule for each device (generating unit for unit commitment and appliances for appliance commitment) over a given period of time based on a number of operational constraints and economical considerations. The formulation of an appliance commitment problem differs from that of the unit commitment in two aspects: (1) the electricity consumption of an appliance is partially driven by random consumer usage (e.g., random hot water consumption), and appliances have fewer power output levels, and (2) the formulation of the TCA constraints is generally more difficult than the formulation of generator constraints (for example, user comfort settings need to be reflected as a set of linear or nonlinear constraints in the optimization problem). Although the algorithm was developed to schedule TCAs, with a little modification, the algorithm can easily be extended to schedule other types of loads, as well as distributed generators or energy storage devices.

Agent-based approaches have been applied to design and operate individual smart appliances [2]. One example of the approach is to let an appliance respond to the deviation of power grid frequency or bid into the market to negotiate the price with others. Compared with the appliance commitment approach, the agent-based approach has the following limitations: (1) the intelligence of each agent on an appliance is very limited, and (2) the coordination among appliances is very difficult. Most importantly, because the appliance commitment algorithm has a scheduling and load coordination process based on price and energy consumption forecasts over a planning horizon (e.g., 24 h), it outperforms the agent-based approach by being more flexible; also, it results in more energy savings.

The formulation of an appliance commitment problem is described here using the thermostatically controlled electric water heater (EWH) load as an example.

Although the scheduling of an EWH has been studied extensively [3–9], and there is a rich literature on how to use the tariff structure to minimize energy payment under different rate structures, the customer comfort level has been rarely discussed. Furthermore, most of the existing scheduling problems were solved by complex nonlinear optimization methods, such as Lagrange methods. A linear approach was proposed by Schweppe et al. [3] and Daryanian et al. [4]. An optimization framework was developed by Mohsenian-Rad and Leon-Garcia [8] to minimize the household's electricity payment by optimally scheduling the operation and energy consumption for each appliance, while considering the waiting time as a comfort setting for the operation of each appliance. A distributed energy resources scheduling algorithm was introduced by Pedrasa et al. [9] to maximize the owner's revenue by considering the deviation from the user-desired temperature as an undelivered energy service. Those approaches have several limitations: (1) in Schweppe et al. [3] and Daryanian et al. [4], a linear thermal model for the water heater was used; (2) no mechanism was used to tackle the prediction errors in price and hot water usage; and

(3) they did not explicitly consider the user choice on thermostat settings as a constraint in terms of a time-varying temperature band.

The appliance commitment approach specifies a time-varying temperature range to reflect consumer choices on the appliances' thermostat settings and their perception of comfort constraints. In addition, the two-step optimization framework proposed can adjust the appliance operation schedule to handle uncertainties that are caused by forecast errors in energy prices and hot water consumption. Moreover, the thermal dynamics of heating and coasting of the EWH load is simulated using physical models; random hot water consumption is modeled with statistical methods. The models are used to predict appliance operation over the scheduling time horizon. User comfort is transformed to a set of linear constraints. The appliance commitment is solved with a multiple-looping algorithm introduced by Lu et al. [10] and enhanced by a novel linear sequential optimization process, which is easy to implement, robust to variations of the forecasted prices, and flexible to handle comfort constraints.

9.2 The Formulation of Appliance Commitment Problem

Household appliances can be classified into three groups: controllable thermostatically controlled appliances (C-TCAs), controllable non-thermostatically controlled appliances (non-TCAs), and non-controllable appliances. Non-controllable appliances cannot be scheduled; therefore, they are modeled by a load profile that is forecasted with historical data. Controllable non-TCAs (such as washers or dryers) are normally discontinuous loads that are straightforward to schedule in an appliance commitment problem. The C-TCA loads, such as HVACs and water heaters, have thermal storage capabilities and are, traditionally, good candidates for consumer-level DR programs. The scheduling of the C-TCA loads has to account for thermal dynamics, random consumption, and consumer comfort. Therefore, the capability of scheduling C-TCA loads is at the core of the appliance commitment algorithm. To simplify our presentation, the scheduling of an EWH load is used as an example to illustrate the formulation of an appliance commitment problem.

To develop an optimal schedule over a time horizon, the appliance commitment requires a price forecast, a range of thermostat settings, and the characteristics of the electric water heater. In addition, the demand for hot water needs to be predicted from time-based heating data.

The primary objective of appliance commitment for a water heater is to minimize the electricity bill over the next 24 h subject to constraints on comfort level and meeting the predicted hot water demand. The comfort zone is defined by the comfort constraints, which reflect the range of the hot water temperature set by consumers. An increase in the comfort zone will provide more flexibility in scheduling, resulting in higher savings. This comfort zone can be time-varying.

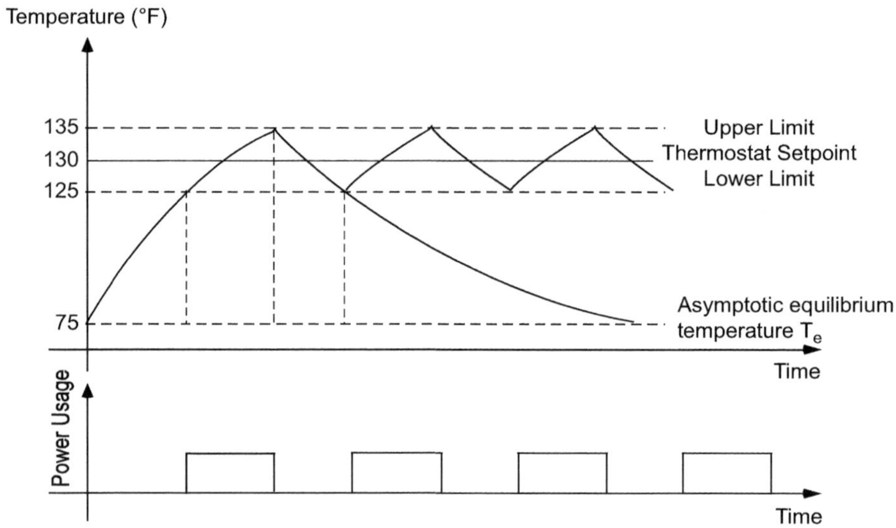

Fig. 9.1 A typical thermal characteristic curve of an EWH load

9.2.1 Thermal Dynamic Modes of EWH

The modeling of an EWH operation requires a thermal dynamic model that describes its heat exchange with the environment and with cold water inflows [11–17]. The thermal model we used is based on an equivalent thermal parameter (ETP) approach [13–17]. Figure 9.1 shows the water temperature inside the EWH tank over time. The rising curves indicate that the EWH is "on," and the falling curves represent the standby (or cooling-down) periods when the heater is "off." Due to the EWH cycles, the water temperature in the tank rises and falls accordingly. The upper and lower limits represent the dead-band of the thermostat around the thermostat set-point.

The thermal dynamic behavior of an EWH can be described by differential equations [13–15]. When the EWH is on over the period $[t_n, t_{n+1}]$ (i.e., $u_n = 1$), its temperature at the time, t_{n+1}, increases to θ_{n+1}, given by

$$\theta_{n+1} = \theta_{\text{en}} + QR - (\theta_{\text{en}} + QR - \theta_n)\exp[-(t_{n+1} - t_n)/(\text{RC})] \tag{9.1}$$

Otherwise, when an EWH is off over the period $[t_n, t_{n+1}]$ (i.e., $u_n = 0$), the temperature at the time, t_{n+1}, drops because of power loss and becomes

$$\theta_{n+1} = \theta_{\text{en}} - (\theta_{\text{en}} - \theta_n)\exp[-(t_{n+1} - t_n)/(\text{RC})] \tag{9.2}$$

Most heaters run because of the household's usage of hot water rather than heater loss. When a household uses hot water, cold water refills at the bottom of tank. The temperature of the water after adding inlet water is given by

$$\theta_n = [\theta_{\text{cur}}(M - d_n) + \theta_{\text{en}} d_n]/M \tag{9.3}$$

Together, Eqs. (9.1), (9.2), and (9.3) describe the thermal dynamics of the EWH, which can be rewritten as a function of thermal parameters, hot water demand, ambient temperature, and on/off status.

$$\theta_{n+1} = f(\theta_n, t_n, Q, C, R, d_n, u_n, \theta_{\text{en}}) \tag{9.4}$$

where

θ_{en}	ambient temperature/cold water inlet temperature (°F)
θ_{n+1}	temperature of hot water at time, t_{n+1} (°F)
θ_n	temperature of hot water at time, t_n (°F)
θ_{cur}	current temperature of hot water (after water draw) (°F)
T_c	temperature threshold used to determine control logic for water heater (°F)
Δt_{on}	duration the water heater needs to be turned on (minutes)
d_n	demand of hot water drawn (or mass of water from cold water inlet) at time, t_n (gallons)
q	thermal energy needed to heat up the water heater (J)
t_n	time at the n-th step (hour)
$\theta_n^{\text{low}}, \theta_n^{\text{up}}$	lower and upper limit of hot water temperature setting (°F)
$\Delta\theta_n$	amount of °C by which the temperature of the added water should be increased
ρ	specific heat of water ($\rho = 4200$ J/kg/°C)
Δt	time elapsed per step (i.e., $\Delta t = t_{n+1} - t_n$) (hour)
C	water heater thermal capacitance (kWh/°C)
M	mass of water in the full tank (gallons)
P_e	power rating of water heater (kW)
Q_{loss}	heat loss of the electric water heater (J)
Q	electric water heater capacity *(kW)*
R	water heater thermal resistance (°C/kW)
m_n	mass of the added water at time t_n (kg) (1 gallon water ≈ 3.785 kg)

9.2.2 The Parameter Estimation of EWH Thermal Model

The EWH thermal model is characterized by thermal resistance and thermal capacitance. These thermal coefficients can be estimated with statistical and regression techniques by fitting the observed performance data, such as temperature and water demands, to Eqs. (9.1, 9.2, 9.3 and 9.4). A detailed discussion on how to use the maximum likelihood method to estimate system parameters can be found in Pahwa and Brice [18].

Table 9.1 Typical values for a water heater

	Q (kW)	R (°C/kW)	C(kWh/°C)	Capacity (gallons)
Values	4	1.52	863.40	50

A typical water heater comes in standard tank sizes that may vary from 30 gallons (113.58 l) to 80 gallons (302.88 l), the most common being 50 gallons (189.25 l) [7]

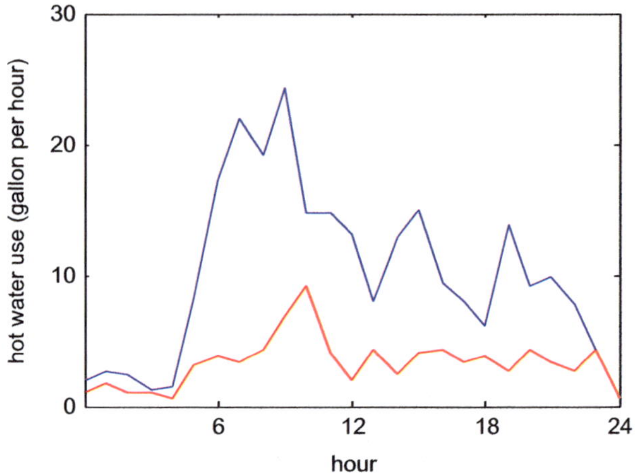

Fig. 9.2 Typical hot water usages of a winter/summer day (blue: winter, red: summer)

An alternative way is to choose the thermal coefficients for water heaters from the 2008 ASHRAE Handbook [19]. For example, typical values for a residential water heater are shown in Table 9.1.

9.2.3 The Random Hot-Water-Consumption Model

Hot water consumption varies from one household to another, depending on the number of users, water-consumption habits (the frequency of hot water use and the duration of the water consumption), and the facilities of a house (bath, shower, etc.). A typical hourly hot water usage in summer and winter is shown in Fig. 9.2.[1]

The historical data of hot water demand can be collected to predict future hot water usage. A flow meter can provide historical data on the time and the volume of water drawn from the water heater. Alternatively, the hourly electricity consumption for an EWH can provide information on the periods during which it is used. Thus, the hot water demand value can be calculated because the electricity usage of a water heater is at a constant rate when it is on. Future demands can be predicted from the

[1]Real measurement data collected in the Pacific Northwest GridWise™ Testbed project [20] were used in our study.

evaluated demand patterns of previous days. Section V has a detailed discussion on EWH load profile forecasting.

9.2.4 *The Comfort Constraints*

Customers may adopt strategies for their water heaters to reduce their energy bills with dynamic pricing. Thus, in contrast to the normal maintenance of the full-storage water heater, one can choose low-cost times for heating water and take advantage of storage capability to have enough hot water to meet the demand at high-cost times. However, it might be necessary to inconvenience the resident to reduce the energy bill. This inconvenience can be that the water temperature is too low at the time of demand, or there is a long waiting time before the temperature is satisfactory. This conflict in two objectives, i.e., minimizing cost and minimizing the negative impact over comfort, has been barely discussed in the literature. Addressing this requires a detailed thermal model of water heaters and an optimal framework that takes into account the operational constraints specified as the comfort level.

Different residences have different preferable temperature ranges for their water heaters. The thermostat settings of water heaters typically vary from 120 to 160 °F. Customers can adjust the setting to reflect the compromise between the comfort level and the energy cost. There is also a safety concern when the temperature of the water heater is too high.[2] In this paper, the acceptable bounds on the temperature, which will be specified by the customer to reflect individual needs, are used as the operational constraint in the scheduling process. Moreover, the result of the schedule as a function of different comfort levels is calculated to offer flexibility to users.

9.3 Solving the Appliance Commitment Problem

Assume that the scheduling of an EWH load is over a 24-h period. The control objective is to minimize the energy bill subject to comfort constraints that reflect a hot water temperature band set by consumers. The hot water consumptions during the scheduling period need to be considered as well. The decision variables are the status of hot water over the next 24 h.

Mathematically, the appliance commitment can be formulated as

[2]Significant scalding risks exist at tap discharge temperatures greater than 120 °F. A few states require set-points in the 120–125 °F range. For example, §704.06 of the Wisconsin State Code requires a residential landlord to set the water heater temperature no higher than 125 °F [7].

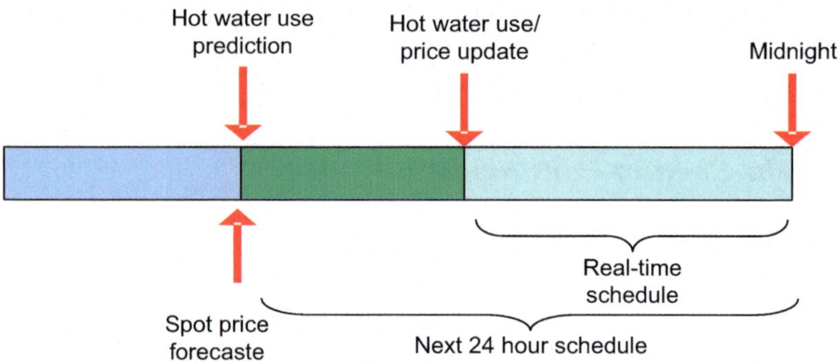

Fig. 9.3 An illustration of the two-step scheduling process

$$\min_{u_n}\left[\sum_{n=1}^{N}(p_n \cdot u_n \cdot P_e \cdot \Delta t)\right] \tag{9.5}$$

subject to the following constraints:

$$\theta_{n+1} = f(\theta_n, t_n, Q, C, R, d_n, u_n, \theta_{\text{en}})\ n = 1 \quad N \tag{9.6}$$

$$\theta_n^{\text{low}} \le \theta_n \le \theta_n^{up}\ n = 1 \quad N \tag{9.7}$$

where

P_{e} power rating of water heater (kW)
u_{n} status of water heater over period [t_n, t_{n+1}] (1-on, 0-off).

The objective in (9.5) is to minimize the cost for energy consumption over the next 24 h. The constraints in (9.6) and (9.7) are the hot water thermal dynamics and the temperature bounds specified by users to reflect their comfort. As the temperature bounds become more restrictive, there is less freedom for the optimal solution. Therefore, the energy consumption scheduling problem is formulated as a nonlinear optimization problem that aims to minimize the electricity payment subject to the user-comfort constraint.

The challenges of the load scheduling are as follows: (1) the model of hot water thermal dynamics is nonlinear, (2) the price forecasts may not be accurate, and (3) the exogenous demand for hot water is stochastic. Therefore, the algorithm has to be flexible and robust to account for the aforementioned nonlinearity and uncertainties.

The model uncertainties can be explicitly handled by applying stochastic dynamic programming [25]. However, this method may be computationally time-consuming. Furthermore, the temperature bounds can be seriously violated for unexpected variations in price or water demand. The above modeling issues motivate us to develop a two-step adjustment process—day-ahead scheduling and real-time adjustment—as shown in Fig. 9.3. First, use the day-ahead prediction of spot prices and hot water usage to solve the deterministic optimal problem over the next 24 h and then adjust the scheduling in real-time based on updated information.

There is a similarity between operating an EWH and an energy storage device. The preheating period of an EWH is similar to the pumping mode of a pumped-storage power station, and the coasting period is similar to its generating mode. Therefore, optimal bidding strategies developed in Lu et al. [10] for a pumped-storage hydro-turbine can be used for applying the preheating and coasting control over an EWH.

To minimize the payment, the preheating period should be done during the least-price hours, and the coasting period should cover the peak-price period. To maximize the comfort level, the EWH should be switched on immediately to restore the temperature after a significant portion of hot water is drawn from the tank, and cold water is refilled. Any delay will lead to a prolonged uncomfortable period or the violation of the temperature-band constraint.

Therefore, the optimization process for the day-ahead scheduling is as follows:

Step 1: Obtain a forecasted, day-ahead, market clearing price (MCP) curve.[3]

Step 2: Based on the forecasted usage of hot water, estimate how long the EWH needs to be "on" in a time interval $[t_O, T]$ to heat up the water in the tank.

The amount of energy needed to restore the energy used is given in [10] as[4]

$$q = \sum_{n=1}^{N} (m_n \cdot \rho \cdot \Delta\theta_n) + Q_{\text{loss}} \tag{9.8}$$

The duration the water heater needs to be turned on to provide q is

$$\Delta t_{\text{on}} = \frac{q}{60P_e} \text{ min} \tag{9.9}$$

This establishes a link between the required operation time of an EWH and the hot water usage of a household.

Step 3: Find the price threshold, P_c, based on how long the EWH needs to be operated.

The estimated price obtained in Step 1 in a time interval $[t_0, T]$ can be sorted ascending to yield a monotonic price curve, w_c. Using Eq. (9.9), the period for the water heater to be on, Δt_{on}, can be calculated. Then, the price threshold, P_c, will be determined from the sorted price curve, w_c, as

$$P_c = w_c(\Delta t_{\text{on}}) \tag{9.10}$$

The control logic for the EWH can be described by

[3]The market price is forecasted on an hourly basis over the next 24 h.

[4]The standby heat loss of commercially available heaters is very small [7].

$$u_n = \begin{cases} 1 & p_n < P_c \\ 0 & p_n \geq P_c \end{cases} \tag{9.11}$$

As such, when the price is lower than P_c, the heater is "on"; otherwise, it is "off."

The control logic in Eq. (9.11), $u_n(n = 1 \cdots N)$, is a solution to the relaxing deterministic optimal problem without considering the comfort constraint, which is given by

$$\min_{u_n} \sum_{n=1}^{N} (p_n \cdot u_n \cdot P_e \cdot \Delta t) \tag{9.12}$$

subject to

$$\theta_{n+1} = f(\theta_n, t_n, Q, C, R, d_n, u_n, \theta_{en}) \; (n = 1 \cdots N) \tag{9.13}$$

Step 4: Check the solution with the status of the EWH determined by Eq. (9.11). Stop if there is no violation of the comfort band. Thus, the total payment, Ψ, is

$$\Psi = \sum_{n=1}^{N} (p_n \cdot u_n \cdot P_e \cdot \Delta t) \tag{9.14}$$

Otherwise, subdivide the time interval into $[t_0, t_1]$ and $[t_1, T]$, where t_1 is the time when the EWH crosses its upper or lower comfort band.

Step 5: After t_1, set the temperature set-point of the water heater to saturated[5] (either as the upper comfort band or as the lower comfort band). For example, at midnight when the price is low, the EWH will be switched on until its temperature reaches the upper bound, i.e., a full tank of hot water is ready for use. After that, the EWH remains off until the temperature of the hot water falls back into the prespecified bound at the time, t_2 (i.e., next significant hot water draw).

Step 6: Set $t_2 = t_0$ and go back to Step 2.

The real-time market prices usually deviate from the day-ahead forecast prices, and the hot water demand may change in real time. The commitment of EWH needs to be adjusted to reflect the real-time changes. Based on the updated hot water usages, the algorithm updates the amount of energy, q, using Eq. (9.8) and then calculates the new operating period $\Delta t_{\text{on}}^{\text{new}}$. With the updated historical price information, re-sort the price curve to yield a new price curve, w_c^{new}. Thus, the updated price threshold, P_c', is

[5]When the EWH attempts to maintain its temperature near the upper or lower temperature limits, the high-frequency switching can occur. To overcome this, a hyperthesis needs to be used (±0.5 °F in this paper). However, depending on the system requirement, a large value may be needed like [7] (±2 °F).

$$P_c' = w_c^{\text{new}}\left(\Delta t_{\text{on}}^{\text{new}}\right) \tag{9.15}$$

9.4 Case Studies

The simulation results are presented in this section. The parameters of the EWH model are given in Table 9.1. The appliance commitment schedule starts from 12 a.m. and runs for 24 h to 12 a.m. the next day.

9.4.1 Day-Ahead Energy Price Forecast

In the simulation, the actual price signal used, p_{n}, was collected in the Pacific Northwest GridWise™ Testbed project [20]. Normally, a day-ahead price forecast curve can be obtained from historical price data using different forecasting methods [21–24].

In our example, the forecasted price signal, p_{f}, is obtained by adding a white noise, ε (which follows a normal distribution), to the actual price signal, as in Eq. (9.16).

$$p_f = p_n + \varepsilon \tag{9.16}$$

A daily mean error of 10% is assumed in this paper, which is consistent with the forecasting accuracy reported by Contreras et al. [21], Hong and Hsiao [22], Nogales et al. [23], and Bohn [24]. The RTP price forecast (red) and the actual price signal (blue) on a typical day are shown in Fig. 9.4.

9.4.2 A State Queueing Model of Water Heater

An SQ model proposed by Lu et al. describes a water heater behavior by tracking the state of the water heater over time (see detailed discussions on this mode in Lu and Chassin [13] and Lu et al. [14]). The evolution of the state of the water heater is mainly driven by the consumer behavior, which is the most important factor that causes the heater to turn on and off in a day. The customer consumptions are classified into two categories: major consumptions and minor consumptions. A major consumption includes behaviors such as washing dishes or taking showers, which usually last for more than 5 min. A minor consumption includes behaviors such as washing hands and washing fruit, which usually last for 1 or 2 min. The probabilities of major and minor hot water consumptions can be calibrated to

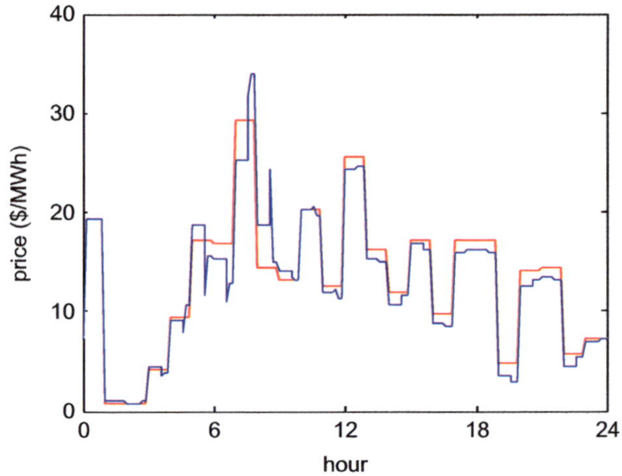

Fig. 9.4 The RTP prices (blue: actual, red: day-ahead forecast)

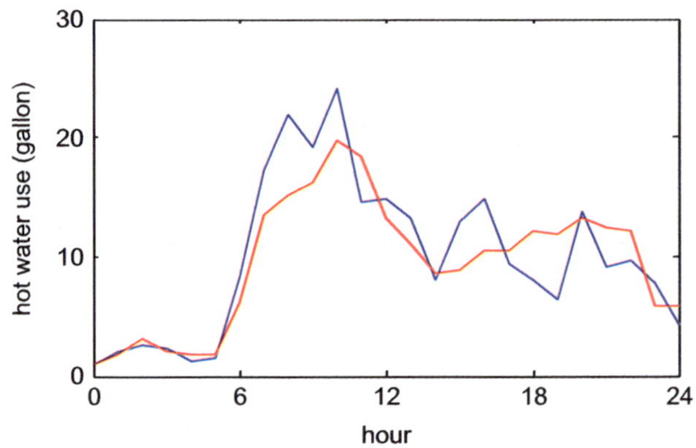

Fig. 9.5 Forecast hot water usage (blue: actual, red: forecast)

minimize the difference between historical demand data and simulation data. Therefore, from these probabilities, simulations can be performed to yield the prediction of the water usage (red) for the next day, as shown in Fig. 9.5.

9.4.3 Transactive Control Strategy

Assume that the upper temperature limit of an EWH is set at 160 °F and the lower temperature limit at 140 °F. If the thermostat of the EWH is not price-sensitive (Scheme A), the water temperature profile (Fig. 9.6 (red)) is simulated using the forecasted hot water usage (the red line in Fig. 9.5). The electricity payment is \$1.018 on this typical day based on the forecasted RTP prices (the red line in Fig. 9.4).

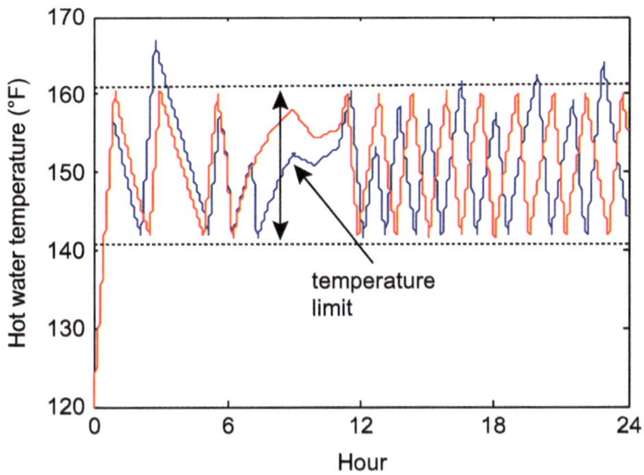

Fig. 9.6 The hot water temperature profiles (Scheme A: no-control [red] and Scheme B: transactive control [blue]) (the upper and lower dot lines represent the temperature limits for Scheme A)

The performance of appliance commitment and the non-price-sensitive scheme is also compared to that of the transactive control strategy applied in the Pacific Northwest GridWise™ Testbed project. A transactive control strategy (Scheme B) applied to a thermostatically controlled HVAC was introduced [2]. The algorithm uses the current indoor temperature to calculate the bid price. After receiving the market clearing price, the adjusted zone set-point is calculated to reset the thermostat. The controller is given a range of daily price variations to adjust the thermostat set-point so that the thermostat reaches its minimum setting when the price is at maximum. If the transactive control scheme is applied to the EWH with the forecast hot water consumption and RTP prices, the electricity payment is \$0.912 on this typical day. The hot water temperature profile is shown in Fig. 9.6 (blue).

The advantage of the transactive control is that it is easy to implement at the appliance level. The disadvantage of the transactive control strategy is that it is a real-time control strategy for individual appliances and cannot optimize the operation of multiple appliances over a time period.

9.4.4 Appliance Commitment

In this subsection, the two-step scheduling process introduced previously is applied on an EWH load to determine the day-ahead schedule and perform real-time adjustments based on the forecasted and actual RTP price and hot water usage, as shown in Figs. 9.4 and 9.5.

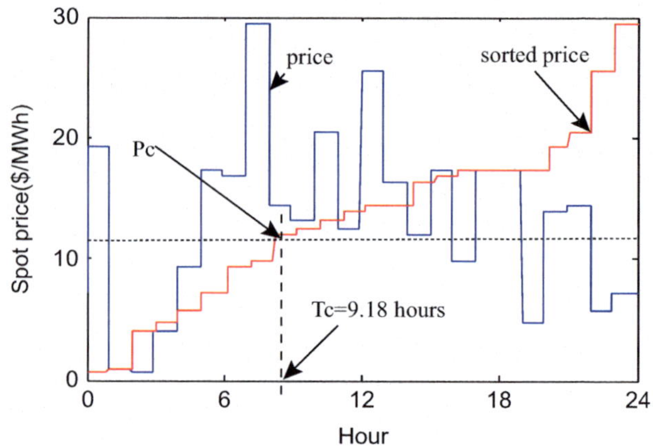

Fig. 9.7 The composite price curve w_c (price threshold, P_c) (blue: day-ahead forecast price, red: sorted day-ahead forecast price)

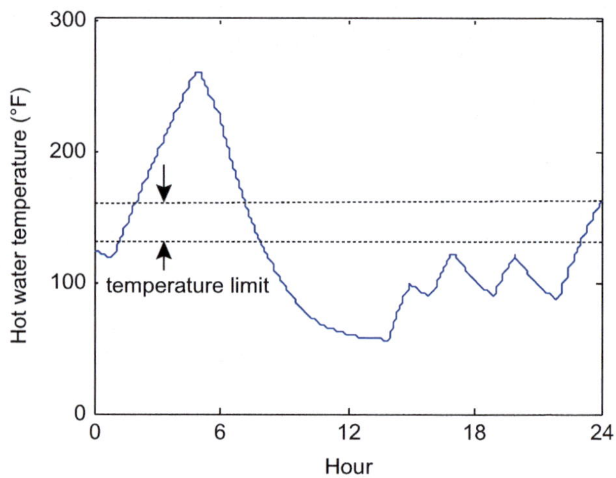

Fig. 9.8 Scheme C: Hot water temperature as a result of the optimal minimum-cost schedule without considering users' comfort (the electricity payment is $0.248)

9.4.4.1 Day-Ahead Schedule

Sort the forecasted RTP prices in ascending order to obtain a composite price curve, w_c, as shown in Fig. 9.7. The price threshold, P_c, is calculated to be 12$/MWh, and the total heater "on" time, T_c, is 9.18 h to meet the heat losses and hot water demand. If comfort constraints are not considered, the operational strategy for the EWH is very simple. When the spot price is less than P_c, the EWH is "on"; otherwise, it is "off." The electricity payment is $0.248, significantly lower than that of the no-control case (Scheme A) or the transactive-control case (Scheme B). However, as shown in Fig. 9.8, the hot water temperature can be higher than 200 °F or lower than 100 °F, which are unacceptable to consumers. Therefore, the comfort constraints reflecting the consumer choices of the hot water temperature band must be included.

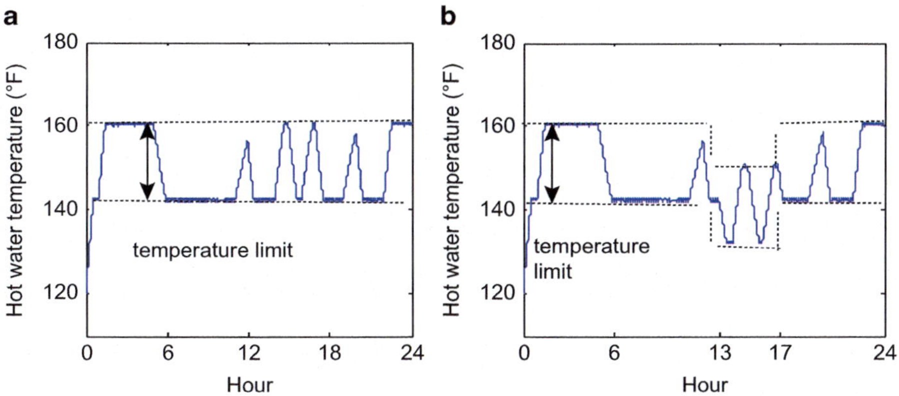

Fig. 9.9 Hot water temperature (°F) as a result of optimal schedule considering users' comfort. (**a**) Scheme D: uniform temperature constraint (the electricity payment is \$0.811); (**b**) Scheme E: time-varying temperature constraint (the electricity payment is 0.798)

In the first case, assume that the temperature band is uniform over the whole scheduling period (24 h),

$$142 \leq \theta_n \leq 160 \text{ when } 0 \leq t_n \leq 24 \tag{9.17}$$

After solving the unconstrained case, check the first violation. As shown in Fig. 9.8, the violation of the upper temperature limit occurs at $t = 1.466$ h; therefore, the whole optimization time period [0, 24] is separated into two: [0, 1.466] and [1.466, 24]. Starting from $t = 1.466$ h, the heater turns off and maintains its temperature until 4.916 h. Then, another unconstrained optimization is carried out over a time period [4.916, 24]; after it is done, the upper and lower temperature limits are checked again. The procedure repeats until no violation of the temperature limits occurs. The resultant electricity payment is \$0.811, and the hot water temperature profile is shown in Fig. 9.9a.

In the second case, the temperature bands change with respect to the time of day:

$$\begin{cases} 132 \leq \theta_n \leq 150 & 13 \leq t_n \leq 17 \\ 142 \leq \theta_n \leq 160 & \text{otherwise} \end{cases} \tag{9.18}$$

The hot water temperature (see Fig. 9.9b) decreases in late afternoon as a result of relaxed constraints in the water temperature. The electricity payment is \$0.798. Compared with \$0.811 in the first case, the economic gain is marginal.

The performance of different control schemes is compared in Table 9.2. Compared with the no-control case (Scheme A), the new schemes (Schemes D and E) can achieve over 20% in savings without compromising users' comfort constraints. Moreover, this flexible framework can accommodate time-varying comfort settings.

Table 9.2 Comparison of performance of different control scheme[a]

Scheme	Min temperature (°F)	Max temperature (°F)	Cost ($)
A	142	160	1.018
B	140	167	0.912
C	66.7	262	0.248
D	142	160	0.811
E	132	160	0.798

[a]scheme A: Conventional thermostat; scheme B: Dynamic price-sensitive thermostat; scheme C: Optimal schedule without considering users' comfort; scheme D: Optimal schedule considering users' comfort (uniform temperature constraint); scheme E: Optimal schedule considering users' comfort (time-varying temperature constraint)

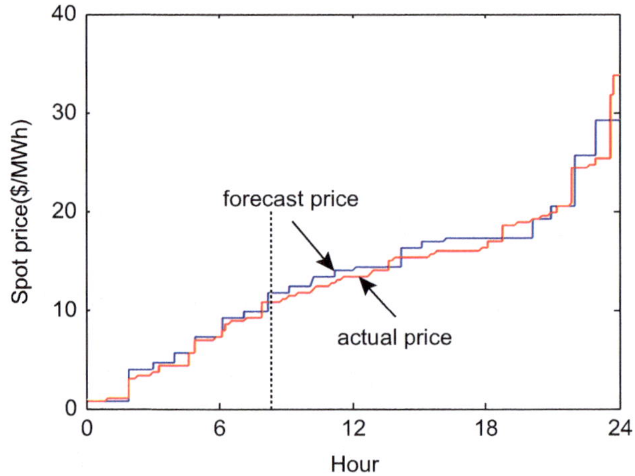

Fig. 9.10 Comparison between the sorted forecast price and sorted actual price (blue: forecast; red: actual)

9.4.4.2 Real-Time Adjustment

The actual price is usually different from the forecast price (see Fig. 9.4). A comparison between the sorted forecast prices and sorted actual prices is shown in Fig. 9.10. If the control logic that is determined based on a day-ahead forecast is used for the actual price signal, it results in about an additional 5 min heating time, and the actual electricity payment is $0.778. The payment difference is small; however, the hot water temperature deviations can be very large because of forecasting errors in price and hot water demand. As shown in Fig. 9.11, the water temperatures (the red line) have crossed the upper and lower temperature limit (140–160 °F).

A real-time adjustment is conducted as described in Sect. 9.3. After the adjustment, the temperature profile (the blue line in Fig. 9.11) is significantly improved; the trade-off is that the cost is slightly higher ($0.831). Note that around 9 a.m., the water temperature dropped below 140 °F even though the water heater was switched on. This was caused by a large hot-water consumption that occurred in that period.

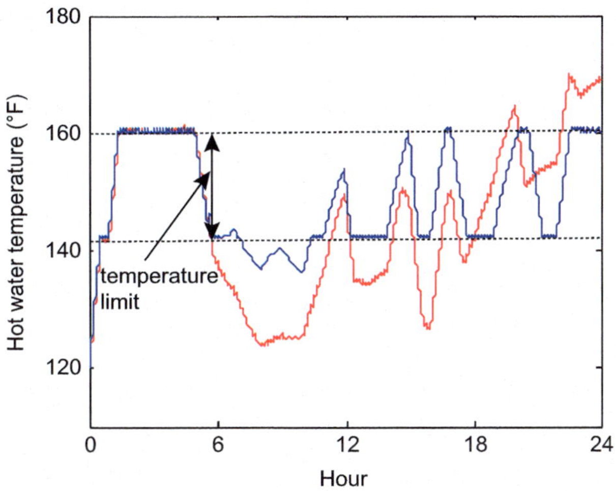

Fig. 9.11 Hot water temperatures (blue: real-time adjustment; red: day-ahead schedule) (the upper and lower dot lines represent the temperature limits)

Table 9.3 Electricity cost as a function of different temperature constraint corresponding to different comfort setting

Case	Lower limit (°F)	Upper limit (°F)	Cost ($)
1	156	165	0.966
2	147	165	0.862
3	138	165	0.763
4	151	160	0.919
5	142	160	0.811
6	133	160	0.712
7	146	155	0.870
8	137	155	0.760
9	128	155	0.663
10	141	150	0.818
11	132	150	0.708
12	123	150	0.612

As demonstrated, the two-step optimization approach can effectively adjust the load schedule to reflect the price changes, thereby accounting for the forecasting error in the price in the problem formulation.

9.4.4.3 The Selection of Comfort Settings

A major benefit of load scheduling is that users can compare the costs/benefits among different comfort settings. Table 9.3 depicts electricity payments associated with different upper and lower temperature settings of the EWH. As the temperature band becomes wider or the temperature band is moved downwards (e.g., from 156–165 °F to 141–150 °F), the cost of electricity is decreased.

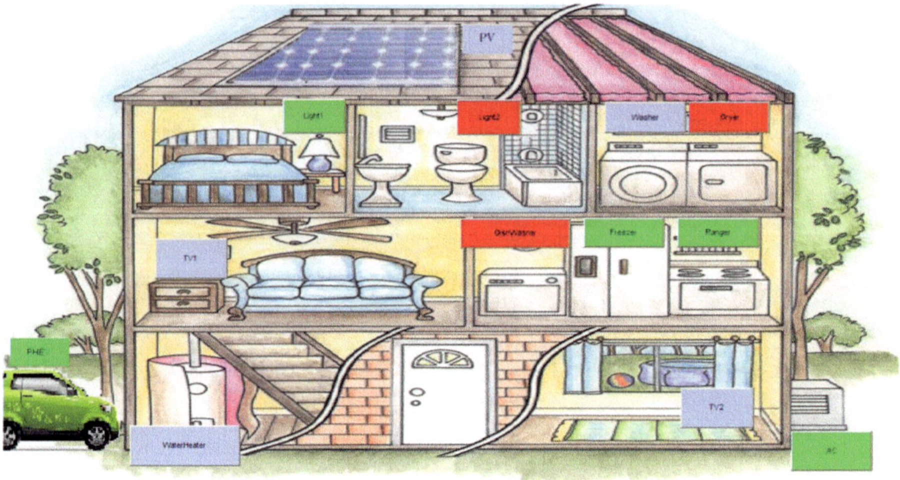

Fig. 9.12 Illustration of a future house with behind-the-meter appliances and DERs

9.5 Future Work

The previous sections describe one example of home energy management system to schedule residential household appliances. As roof-top solar resources together with demand response and other distributed devices, collectively referred to as distributed energy resources – DERs, proliferate in the low-voltage distribution system – see Fig. 9.12, a single behind-the-meter energy management gateway has the potential to enhance situational awareness, maximize financial benefits, and enable resiliency by controlling all of them in a systemic and cohesive manner. However, achieving this vision requires a control platform that offers intelligent, prognostic scheduling of DER devices, flexible, interoperable interfaces to end-point devices that can enable broad-based adoption and support for grid-connected and off-grid operation to maximize grid resiliency and reliability.

The coordination among these behind-the-meter assets provides a viable solution to both the grid and the end users. The unified control platform can satisfy customer requirements along multiple dimensions such as usability, performance, and privacy concerns. For example, it can change the load shape following the movement of the real-time energy price or to maximize a user-defined utility function (e.g., minimize reverse power flow, peak shaving, and energy arbitrage). The optimal coordination of these assets can fully exploit the flexibility of a house load demand and DERs and provide multiple services to the grid operators. With an advanced control, the whole house can act like a virtual power plant and thus is capable to follow the dispatch instructions from system operators. To this end, flexible load demand plays a critical role in the coordination as it is still the most cost-effective resource to schedule and control.

On the other hand, the unified control involving different asserts increases its complexity of the problem and it may be challenging to seek a solution. This requires the details of load characteristics and load flexibility to be modeled. Additionally, the interactions between different resources may introduce nonlinear dynamics and non-convex constraints. To better assess the economics and feasibility of the solution, the uncertainty from roof-top solar and user behaviors also need to be considered when designing such a schedule. Despite all of the difficulties, more solutions will be developed as the researchers are inspired by the full potentials which a smart house could deliver to both the grid and the customers.

9.6 Conclusions

This chapter illustrates the formulation of an appliance commitment problem and presents a two-step linear-sequential-optimization-enhanced, multi-loop algorithm to solve the problem. The features of the proposed approach are summarized as follows:

(a) The appliance commitment algorithm provides a systematic way to formulate an optimal load scheduling process by modeling the appliance thermodynamic process and explicitly considering the user comfort setting as a constraint. The solution achieves an optimal balance between energy cost and user comfort level.
(b) The two-step appliance commitment algorithm provides adjustments to the optimal schedule to account for uncertainties introduced by forecast errors in prices and consumer consumptions.
(c) The algorithm is fast and robust. The central processing unit (CPU) time of the optimization results solved for a 24-h EWH schedule on a PC (Intel Duo CPU@3.00GHz) is 11 s on average. Because the linear-sequential-optimization-enhanced, multi-loop algorithm is fundamentally an exhaustive search algorithm, the solution is optimal and always solvable.

The algorithm can be used in home/building energy-management systems to help household owners or building managers to automatically create optimal load operation schedules based on different cost and comfort settings and compare cost/benefits among schedules.

References

1. Wood, A. J., & Wollenberg, B. F. (1996). *Power generation operation and control*. New York: Wiley.
2. Katipamula, S., Chassin, D. P., Hatley, D. D., Pratt, R. G., & Hammerstrom, D. J. (2006). *Transactive controls: A market-based GridWiseTM controls for building systems*. Richland, WA: Pacific Northwest National Laboratory.

3. Schweppe, F. C., Daryanian, B., & Tabors, R. D. (1989). Algorithms for a spot Price responding residential load controller. *IEEE Transactions on Power Systems, 4*, 507–516.
4. Daryanian, B., Bohn, R. E., & Tabors, R. D. (1989). Optimal demand-side response to electricity spot prices for storage-type customers. *IEEE Transactions on Power Systems, 4*, 897–903.
5. Lee, S. H., & Wilkins, C. L. (1983). A practical approach to appliance load control analysis: A water heater case study. *IEEE Transactions on Power Apparatus and Systems, PAS-102*, 1007–1012.
6. Ericson, T. (2009). Direct load control of residential water heaters. *Energy Policy, 37*(9), 3502–3512.
7. Goh, C. H. K., & Apt, J. *Consumer strategies for controlling electric water heaters under dynamic pricing*. Carnegie Mellon Electricity Industry Center Working Paper CEIC-04-02.
8. Mohsenian-Rad, A.-H., & Leon-Garcia, A. (2010). Optimal residential load control with price prediction in real-time electricity pricing environments. *IEEE Transactions on Smart Grid, 1*(2), 120–133.
9. Pedrasa, M. A. A., Spooner, T. D., & MacGill, I. F. (2010). Coordinated scheduling of residential distributed energy resources to optimize smart home energy services. *IEEE Transactions on Smart Grid, 1*(2), 134–143.
10. Lu, N., Chow, J. H., & Desrochers, A. A. (2004). Pumped-storage hydro-turbine bidding strategies in a competitive electricity market. *IEEE Transactions on Power Systems, 19*, 834–842.
11. van Tonder, J. C., & Lane, I. E. (1996). A load model to support demand management decisions on domestic storage water heater control strategy. *IEEE Transactions on Power Systems, 11*(4), 1844–1849.
12. Nehrir, M. H., Dolan, P. S., Gerez, V., & Jameson, W. J. (1995). Development and validation of a physically-based computer model for predicting winter electric heating loads. *IEEE Transactions on Power Systems, 10*(1), 266–272.
13. Lu, N., & Chassin, D. P. (2004). A state queueing model of thermostatically controlled appliances. *IEEE Transactions on Power Systems, 19*(3), 1666–1673.
14. Lu, N., Chassin, D. P., & Widergren, S. E. (2005). Modeling uncertainties in aggregated thermostatically controlled loads. *IEEE Transactions on Power Systems, 20*(2), 725–733.
15. Katipamula, S., & Lu, N. (2006). Evaluation of residential HVAC control strategies for demand response programs. *ASHRAE Transactions, 112*(1), 1–12.
16. Pratt, R. G., & Taylor, Z. T. (1994). *Development and testing of an equivalent thermal parameter model of commercial buildings from time-series end-use data*. Richland, WA: Pacific Northwest Laboratory.
17. Taylor, Z. T., & Pratt, R. G. (1988). The effects of model simplifications on equivalent thermal parameters calculated from hourly building performance data. *Proceedings of the 1988 ACEEE Summer Study on Energy Efficiency in Buildings*, (pp. 10.268–10.285).
18. Pahwa, A., & Brice, C. W. (1985). Modeling and system identification of residential air conditioning load. *IEEE Transactions on Power Apparatus and Systems, pas-104*(6), 1418–1425.
19. 2008 ASHRAE Handbook—HVAC Systems and Equipment, American Society of Heating, Refrigerating and Air-Conditioning Engineers, Inc., 2008. ISBN 978-1-933742-33-5. Retrieved February 16, 2011, from http://www.knovel.com/web/portal/browse/display?_EXT_KNOVEL_DISPLAY_bookid=2396
20. Hammerstrom, D. J., Brous, J., Carlon, T. A., Chassin, D. P., Eustis, C., Horst, G. R., Järvegren, O. M., Kajfasz, R., Marek, W., Michie, P., Munson, R. L., Oliver, T., & Pratt, R. G. (2007). *Pacific northwest GridWise™ Testbed projects: Part 2. Grid friendly™ appliance project* (Vol. PNNL-17079). Richland, WA: Pacific Northwest National Laboratory.
21. Contreras, J., Espínola, R., Nogales, F. J., & Conejo, A. J. (2003). ARIMA models to predict next-day electricity prices. *IEEE Transactions on Power Systems, 18*(3), 1014–1020.

22. Hong, Y.-Y., & Hsiao, C.-Y. (2002). Locational marginal price forecasting in deregulated electricity markets using artificial intelligence. *IEEE Proceedings—Generation, Transmission and Distribution, 149*(5), 621–626.
23. Nogales, F. J., Contreras, J., Conejo, A. J., & Espínola, R. (2002). Forecasting next-day electricity prices by time series models. *IEEE Transactions on Power Systems, 17*(2), 342–348.
24. Bohn, R. E. (1982). *Spot Pricing of Public Utility Services*, Ph.D. dissertation, MIT, 1982. Also available as MIT Energy Laboratory Technical Report, MIT-EL 82-031, 1982.
25. Tuohy, A., Meibom, P., Denny, E., & O'Malley, M. (2009). Unit commitment for systems with significant wind penetration. *IEEE Transactions on Power Systems, 24*(2), 592–601.

Index

P. Du et al., *Demand Response in Smart Grids*,
https://doi.org/10.1007/978-3-030-19769-8

Printed by Printforce, the Netherlands